Make:

3 MODES OF MAKING

Make:

3 MODES OF MAKING

MATT ZIGLER

Foreword by Dale Dougherty

Make:
3 MODES OF MAKING

By Matt Zigler

ISBN: 978-1-68045-799-5

December 2023: First Edition

See www.oreilly.com/catalog/errata.csp?isbn=9781680457995 for release details.

MAKE: BOOKS
President Dale Dougherty
Creative Director Juliann Brown
Editor Kevin Toyama
Copyeditor Ann Martin Rolke
Proofreader Emily Gertz
Illustrations Matt Zigler

Make: Community is a growing, global association of makers who are shaping the future of education and democratizing innovation. Through Make: magazine, 200+ annual Maker Faires, Make: books, and more, we share the know-how of makers and promote the practice of making in schools, libraries, and homes.
Make: books may be purchased for educational, business, or sales promotional use. Online editions are also available for most titles. For more information, contact our corporate/institutional sales department at 800-998-9938.

Make Community, LLC
150 Todd Road, Suite 100
Santa Rosa, California 95407
www.make.co

CONTENTS

CONTENTS

FOREWORD
by Dale Dougherty

If there were a central tenet of Maker Education, it would be that students are empowered to do real projects based on their own ideas and interests — to be makers. A maker starts with an idea, develops it as a project, and then turns it into something real, something that demonstrates what the idea is and what it can do.

That's a much more radical proposition for education than it might sound at first.

Most students in traditional schools don't get to do projects. Instead, they are asked to sit and listen, raise their hand if they have a question, and take quizzes and tests. Students are not given the opportunity to create or build anything. Most teachers are tasked with teaching a subject — such as English, history, science, or math — and present information, typically from a set curriculum or textbook, and then test students to see what they recall.

However, this is not how we teach art, sports, or music. (We could teach them this way but we don't — because it doesn't work.) Students learn art by using different materials. They learn music by learning to play an instrument, and learn sports by playing games. They are encouraged to practice and play. They learn from experience, not theory.

These are all examples of learning by doing, and the focus on doing is what distinguishes maker education. It is about what you can do with what you know, and I don't think this is unique to learning art, music, and sports. A student can "do" history, "do" science, and "do" English in many creative ways, but that seems unexplored.

I happen to think that the reason science is not a popular subject in schools has more to do with the way that it is taught than the subject itself. What does it mean to "do" science? That's why I think science teachers could learn more from art teachers about how to teach their subject.

I imagine that an art teacher finds some students believe they are not good at art, while others feel confident about doing a particular kind of art because they have already shown some proficiency for it but may be afraid to explore new tools or techniques.

When a student believes that "I'm not an artist," the art teacher must coax them just to try to draw or paint. "Don't worry whether it's any good, that doesn't matter," the teacher might say. "Just do something." Getting a student to be open to the experience of doing — however it turns out — is as important as teaching a student how to use a particular material or technique. Once I saw a teacher substitute a reluctant student's blank piece of paper with a grocery bag, and encourage the student to just scribble all over it. The switch worked: the reluctant student began to draw.

Teachers may also have to coax some students out of their comfort zones, and encourage them to explore new techniques and new materials without judging the outcome in advance. It might be necessary to acknowledge the student's discomfort while encouraging them to overcome it. With both groups, the teacher's focus is on helping students engage in a process, a creative process that's also a learning process. Teaching students how to acquire knowledge through experience is what many art teachers are trained to do well.

A makerspace has a lot in common with an art studio. Both are places that provide people with tools, workspace, and materials. They are places to work on projects. They are places where people can learn about new technologies and techniques, as well as learn from and be inspired by the work of others.

Matt Zigler is an art teacher who also runs his school's makerspace. When he gave a presentation at our Make: Education Forum in 2022, I was struck by his key insight that there are three modes of making and all three are ways that engage students as learners.

So, I asked him to develop his presentation into a book for fellow educators that would help them with the challenge of applying process-oriented learning — making — to all sorts of subjects beyond art or sports or music. It would help them also meet the challenge of coaxing students to engage in developing projects of their own.

Imitation, modification, and innovation is how we acquire and practice real-world skills. The learning process — and learning how to learn — is something we can use to solve problems in a variety of different contexts.

Take cooking. We may first learn to cook by following recipes. Then we learn how to make adjustments to recipes that reflect our own tastes. Eventually, we may be able to cook well without recipes, or may even create recipes of our own. We may seek out inspiration from others, too: One

FOREWORD

reason cooking shows are perennially popular is that watching other, better cooks is a great way to learn and become better cooks ourselves.

Surprisingly, learning to write computer programs is not that different from learning to cook. We may try using short examples of code — programs — written by others, and then rewrite the program slightly to make the computer do something a little different. Practice this regularly, the same way you'd practice playing an instrument, and gradually the ability to tweak a program will expand into the ability to write your own program from scratch.

I said earlier that getting students to do their own projects was a radical proposition. It is not something many teachers have been trained to do. It requires a different approach, which might be called facilitation, or what I prefer: coaching. A coach guides practices and their goal is to improve the performance of the student. Like the art teacher, they may have to coax some students, while others will be ready to try new things.

Coaching students who are doing individual or group projects can seem daunting, but it can also be rewarding when it succeeds at getting students engaged with learning. One of the most satisfying comments that I've heard from a maker educator is: "This is why I wanted to be a teacher in the first place."

It does require more of the teacher and even more of the student, but that's also the reward. When both teachers and students are engaged in a creative, learning process, making becomes meaningful — and that brings out the best in everyone.

DALE DOUGHERTY believes that all of us are makers and he is a champion of the people and projects that form the Maker Movement. He founded *Make:* magazine in 2005, and first used the term "makers" to describe people who enjoyed "hands-on" work and play. He started Maker Faire in the San Francisco Bay Area in 2006. He is President of Make Community, LLC, which produces *Make:* magazine and Maker Faire.

INTRODUCTION

When I was 6 years old, my family moved from the suburbs of Washington, DC, to a small farm an hour outside of the city. My mom and dad appreciated the connection to their agricultural roots. My brother gravitated to the scientific elements. I got a sketchbook and started drawing pictures of birds, flowers, and other natural subjects. I had lots of LEGO bricks, army men, and various other toys, but when I thought of myself, it was as someone who liked to make art.

During high school, I had two very different experiences with art and education. I took the various drawing and painting classes that were offered at my public school, and at the same time I started taking weekend classes with a well-known local artist. For several years my weekends were spent in his garage studio painting still life, going out to sketch or paint landscapes and cityscapes, looking at the work of other artists, and getting my work critiqued on an ongoing basis.

This experience was significantly different from my art class experiences in public school and introduced me to the idea of studio practice. There were no grades, rubrics, deadlines, or assignments other than to sketch and paint. I was challenged with different techniques, media, and subjects (like when he brought his pet rooster into the studio for us to draw). I attempted to do the best work I could, with no formal assessment other than the discussion and critique from the teacher and my peers. I would make some sketches through the week and then come back for more, only this time in a pasture with cows milling about. Through this practice I became a better painter.

The art created in my school classes was very assignment based. There were projects on color mixing, contour line drawing, and shading. I remember the shading assignment was a rendering of a clump of ribbons in shades of gray. Each of us did very much the same drawing, with varying skill levels being the differentiator. We learned technical skills of art, rather than the skills of being an artist.

These two experiences of learning art and creativity, happening concurrently, had a profound effect on me. It became clear that making art did not necessarily mean being creative, and vice versa. It's possible to be surrounded with creative tools and materials, and not be creative at all.

These experiences and others prompted me to embrace a studio practice rather than an academic art tradition. My mentor commented that if I made 100 paintings, I would learn something about painting. This statement inspired me to paint nearly every day in the park behind the school, and it was probably the most artistically productive time of my life. I decided to be an art teacher to inspire creativity in others, but it wasn't until I became a maker educator that I truly felt this was possible.

After 20 years of teaching in the art studio and the makerspace, I am still not sure when I am getting it right. Every year I tweak and change how I work with students, how I give them feedback, and how I assess their efforts. This book is the result of many years of trying to build the sort of environment I found in my weekend art classes within a traditional school setting. It is a collection of ideas, strategies, and tools that I have found to be useful in the challenging task of empowering young adults to become creative problem-solvers that can change the world.

1

THREE MODES OF MAKING

In *The Art Spirit*, Robert Henri, the influential art teacher at The Art Students League during the early 20th century, wrote:

> ***Art when really understood is the province of every human being. It is simply a question of doing things, anything, well. It is not an outside, extra thing. When the artist is alive in any person, whatever his kind of work may be, he becomes an inventive, searching, daring, self-expressing creature.... Where those who are not artists are trying to close the book, he opens it, shows there are still more pages possible.***

This book is about making, creativity, and the ways that they take place. Kids are often labeled as creative or not, as if there is one way to be creative. For much of my life, I considered myself an artist, and even more specific than that, a painter. Over time my identity expanded to other forms of art, and once I stepped into the makerspace, technological tools such as CAD, coding, and electronics joined my toolbox. I use all of these accumulated tools and skills in a variety of ways, but over the years I have come to identify three different modes of making into which all of my projects inevitably fall: imitation, modification, and innovation.

As a maker educator, these three modes of making have become central to how I think about the teaching I do, the projects I facilitate, and the way I work with students and teachers. Over the last two decades, I have been increasing the intentionality of how I teach these three modes of making, and this book will lay out the framework that I use to teach purposeful maker classes in high school.

Teaching Creativity

Teaching high school art in a studio practice method is easier said than done. Students who take weekend art courses have self-selected because of a strong interest in art. In many schools, especially small schools, art may be one of the few electives offered, and students in those classes may range from avid artists to reluctant or even hostile participants. Even if the class is full of motivated young artists, there are still grading requirements to fulfill and curriculum standards to uphold. Students who feel they aren't creative or "can't draw a straight line" feel like they don't belong. Art may be the only subject where it is assumed you need to know the content before you take the class, not after.

Every year students would take my art classes and say, "I don't even know how to draw, why do I have to take this class?" I would point out that most students don't know how to calculate the area of a rectangle BEFORE taking geometry, or how to dissect a frog BEFORE taking biology. The truth is, however, that all those students had taken classes called "art" for years before, in elementary and middle school, and had already formed their opinions about what defines art. Half of my job was to disrupt that thinking so they would be open-minded enough to give it a try. I wanted my classes to be about creativity, not just art.

When you teach engaged and self-motivated students, you can make inspirational statements and let the students leap to understanding. When my mentor suggested that I make 100 paintings, he was trying to motivate me to establish a practice. He was right: I certainly did learn something about painting, but though the challenge was inspiring, it might not exactly be called "teaching." James Elkins, in the book *Why Art Cannot Be Taught*, describes a critical element of what we call teaching as "intentionality." We can learn things without a teacher through direct experience, research, trial and error, and other ways. Teaching implies that we intend for the student to gain some knowledge through our words, actions, or lessons. Learning may happen randomly or unpredictably, but teaching (even if I am wrong or misguided) must be intentional.

When I became an art teacher, I wrestled with this issue of intentionality. Why are some students able to pick up skills so much quicker than others, and how do I help those that are struggling? What do I do for the kids who don't think they are creative/artistic/able to draw a straight line? What about the kids who don't want to be in my class in the first place? There will always be variability in the experience level and motivation of students, but I as the teacher should be able to create an environment where students can reach their highest potential.

There are clearly techniques that can be taught. I can teach a student the step-by-step process for one-point perspective drawing, and I can teach the proportions of the face, so my students don't have to work it out for themselves. I can teach them about the artists and art movements that came before, but how do I teach them to do something new to express themselves? If that can't be taught with, as Elkins says, *intentionality*, what can I do for my students to help them grow their creative potential? How can I give them that studio experience of figuring things out for themselves by exploring and taking risks without it being random or haphazard? For the many years that I taught art, I sometimes thought I accomplished this, while other times felt like it was impossible.

After 15 years of teaching in the art studio, I moved to teaching in the makerspace. I suddenly had a degree of freedom in teaching creativity that I had never felt before. Art education has a weight of expectations, history, and methods that are hard to shift. The makerspace is a new frontier in terms of pedagogy, and I loved the opportunity to think about how I worked with students and what the goals of various activities were. I thought back to memorable moments and impactful lessons I had been a part of, and tried to look for common themes. I thought about how to structure my class and the activities I do with teachers in their content areas.

I started with analyzing "how to draw" books, paint-by-numbers, and step-by-step tutorials, such as LEGO kit instructions. What did I learn from these experiences of imitating work done by others? These activities are seen as less significant than "higher level" creative projects. Yet, they ranked among some of the most memorable learning experiences in my artistic life, because I gained specific skills and techniques with tools and materials.

Drawing or working from life seemed different than step-by-step tutorials. There was an element of copying in this type of work, but every artist edits what they see, simplifies, improves, and adds to it. When I painted in the park behind the school, I gradually came to know that place. Each painting was, in some ways, based on the previous ones, improving and modifying the prior experience and taking more ownership. Creating art by taking something that exists and editing, improving, and personalizing it is a significant skill separate from the ability to follow an established procedure.

Finally, there were experiences where I felt like I was breaking new ground and truly expressing my individual vision. In these cases, the process was one of exploration and experimentation. It didn't matter that my ideas drew from things I had seen and artists I had studied. The skills I put into action were those needed for innovative work in general. The feeling of discovery that you get when pursuing something new is fuel for the creative engine within all of us. Everyone is born with the ability to be creative, but not everyone uses their creativity. As with anything that can be used, it takes practice to use it well.

Why Do High Schoolers Become Less Creative?

In 8th or 9th grade, the traditional curriculum shifts, and little silos pop up. STEM becomes science in period one, math in period three, and perhaps some kind of robotics or engineering in an elective slot. The art and creativity built into many elementary and middle school technology classes get pushed into a traditional arts track, leaving the technology behind. These subjects, not to mention creativity and design, become independent of each other, and as a result, students compartmentalize them. I can't tell you how often students don't realize that all the geometry they learned can be applied to real materials like wood, fabric, and metal.

What's worse is that these classes focus on facts and the subject's history, not on encouraging students to gain the thinking skills needed to do math or science in the real world. I once asked my father and brother, who both earned their PhDs in biology at Duke, if they could remember an inspiring time from their high school science classes that led them down their chosen paths. Neither could come up with one. Both were inspired to become scientists by experiences from outside of school, like my weekend art classes.

The reality is most of the accumulated knowledge of these subjects can be accessed from a device that fits in your pocket. Skilled doctors and engineers only hold some of the relevant information in their heads; they know how to find the information they need for the problem they are addressing. We don't need a curriculum based on information; we need one based on experience.

In *Visual Thinking*, Temple Grandin points out that traditional education today focuses almost entirely on verbal thinking. Students write essays and answer multiple-choice questions to show their understanding, rarely drawing pictures and diagrams. Visual thinkers imagine pictures, diagrams, and spatial relationships and have historically been builders, engineers, and inventors. Grandin says that algebra, which is almost entirely abstract and has very little to visualize, screens out many visual thinkers from more advanced STEM subjects. Hands-on courses such as shop class and home economics have almost entirely disappeared in the push for all students to seek out a college degree.

The Role of Maker Education in High Schools

It doesn't have to be this way. Imagine a science class where the students start with the beginning of the scientific method—a question or a hypothesis. To answer their question, they learn how a scientist would go through research, experimentation, and testing of variables. They explore existing data, current and past theories, how to use data, design experiments, and present their findings. They would not all learn the same information, true. Still, they would leave that class understanding how to find the information they might need to answer any questions, as well as how to think like a scientist, a mathematician, a designer, or an engineer.

Some schools do exactly this kind of teaching and learning, but they serve a fraction of a fraction of students. Ninety percent of American students attend public schools with traditional curriculums guided by federal and state standards, approved textbooks, and standardized teacher evaluation criteria. If you don't "cover the content," you won't be a teacher for long, and though most private and charter schools have more freedom, they still teach these subjects in traditional ways because that is what the parents and administrators expect. Changing how core subjects like math and science are taught on a large scale may not be possible within our lifetimes.

Makerspaces and high school maker courses provide opportunities to offer just these sorts of experiences, though, allowing students to learn how to think like engineers, inventors, designers, and builders and offer them the opportunities to apply those skills to real projects. Makerspaces seem "innovative" and don't carry the weight of expectations of the science lab, math class, or even the art studio. They have been more prominent in private schools but are finding their way into public schools more often.

If they don't play a meaningful role in high school education, or if they are sidelined as robotics and engineering classrooms alone, they won't contribute what is sorely needed in school: excitement about learning how to solve real problems in creative ways.

Coming up with a meaningful maker education curriculum for high school and beyond is vital for many reasons. Gaining experience in the creative process is an important life skill in any meaningful career. In 2010, IBM conducted a global CEO study, surveying over 1,500 CEOs in 60 countries. These CEOs identified creativity as the most important leadership quality.

The study report highlighted standout leaders as those who promote experimentation and innovation at all levels of their organization, not just the top. Creative leaders were more likely to take calculated risks, try new ideas, and look to constantly improve how they do business. The word *innovation* features prominently in every subsequent report, indicating that businesses can no longer get by with minor tweaks and improvements. The best jobs in today's economies want people with experience practicing creativity in strategic and applicable ways.

There are many fantastic resources for teaching making and design to younger kids. When I go to conferences, most project sessions are for elementary and middle school students. I usually attend those sessions anyway for the sheer creativity and joy in making that they exhibit, and I often find that with a bit of tweaking, there are ways to apply them to my older students. Still, I often wish there were more offerings for high school students beyond the standard technology electives.

In this book, I lay out a framework for maker courses in upper grades that intentionally teach students various creative process skills through the making of objects in three different modes: imitation, modification, and innovation. Students can employ these modes as needed for a variety of creative tasks. By focusing on teaching skill development rather than how to build specific objects, students will gain experience with techniques to improve and enhance their creative skills and learn how to apply them to challenges in a wide variety of fields. By building strong maker education programs in high schools and colleges, we can work towards fostering the creative problem solvers that the world needs for the future.

2

IMITATION, MODIFICATION & INNOVATION

For the past several years as a maker educator, I have been using the concepts of imitation, modification, and innovation as the structure for how I think about the curriculum I teach, how I assess students, and most importantly, what I want them to get from their experiences in the makerspace. Expertise and mastery over the skills involved in these three modes of making are universally applicable in any pursuit that involves creative thinking and problem-solving.

It's important to think of these three modes as equally essential and connected. There is often a tendency to imagine that imitation is only valuable to pursue modification, which leads us to innovation, and therefore imitation is in some way a lower-level skill. Many times at conferences and on social media, I have heard maker educators dismiss "kit projects" as lesser than other projects, but this shows a focus on product over process. When we focus on the process of imitation, we can look at the various STEM kits, tutorials, and pre-packaged projects out there and evaluate them on how well students can practice the skills of imitation.

Imitation: Working from a set of steps to build foundational skills, basic knowledge, and hands-on experience.

Imitation, when done well, is about improving technique and concrete skill acquisition. It is a way of learning from what has come before and can be done accidentally or intentionally. In chapter 3, I will delve into many skills students can build through quality imitation projects.

Modification can also get a bad reputation in that it is considered a surface-level activity at best and, at worst, theft or copyright infringement. Modification can range from engraving your name onto a water bottle to altering existing code, or taking a design intended for one purpose and applying it to another. It is taking something that already exists and modifying it in a way that suits you or another better. Taking some level of ownership of the object and design is a crucial factor. Open-source design and Creative Commons licensing are both based on the idea that we may take designs and tweak, alter, and improve them while recognizing the chain of creators that led up to our contribution.

Modification: Remixing existing ideas and objects with tools, materials, and techniques with which you have become familiar.

The result of a modification project doesn't significantly change the function or application of the original. The engraved water bottle is still a water bottle, and the edited code doesn't serve a radically different function; they are both improvements over the original while maintaining the best qualities of the original object. As with imitation, many skills can be learned through modification, and chapter 4 will discuss those in more detail.

The stated goal of many makerspaces is innovation. The name of the makerspace where I work has innovation as part of its acronym. Innovation is difficult to define, and often it is easier to point at it than to explain what it is.

Innovation: Combining previous experiences and skills to create something "new" and bringing those ideas into the world.

I put the word new in quotation marks because there are no genuinely new ideas, in the sense that they come out of nowhere. New ideas are significant steps taken beyond existing ideas, pushing them into uncharted territory.

Innovation happens through exploratory processes such as prototyping and iteration. Innovators are feeling their way towards a stated goal, relying on skills and experiences they already possess, but also figuring out things on the fly as needed.

An innovative project often begins as a form of modification but goes just a little further. As students make these moves into new territory, they will often need to rely on imitation to pick up the skills required to make that next step. Therefore, innovation cannot be separated from modification and imitation because it relies on them to reach the goals of the innovator.

This method of making can be the most daunting for teachers in the makerspace. Students will inevitably come up with ideas and goals beyond the teacher's skills, but this is how students learn creative problem-solving, and the teacher is there to help guide them through the process, even though they may not be able to help them with the product. I delve into the skills involved in innovation in chapter 5 and see how this takes place in a class in chapter 6.

A long time ago, I gave up on being the "sage on the stage." Every year my students want to pursue projects that are beyond my knowledge and abilities. I let them know ahead of time that they will need to pursue their own sources of information and knowledge. I can help them ask questions, I can help them seek out sources, and I can help them think through their problems, but I can't tell them how to do it. Rather than feeling unqualified or powerless, I feel proud that they are willing to risk an incomplete or failed project.

None of the skills involved in these three modes of making are new and groundbreaking. Attention to detail, finding personal meaning in projects, and looking at the big picture are all things that we do in various ways at a very young age. The ability to do them with purpose and to learn from the intentional application of these skills is what students can get from a maker curriculum based on imitation, modification, and innovation.

Project Examples

Imitation Projects

Up to this point, I have talked a lot about imitation, modification, and innovation as concepts. To make this a worthwhile structure for teaching intentionally in the makerspace, there needs to be plenty of concrete information to implement these concepts. What are actual projects that fit into these categories?

Imitation projects are, by nature, the easiest ones to list as they are also the easiest ones to find. They range from fully contained kits with all the parts needed to complete the project to instruction sets that require the builder to procure the necessary parts and tools. In today's digital fabrication age, many projects fall in between with part files that can be downloaded and fabricated, then assembled based on the provided instructions.

There are an uncountable number of kit projects that you can purchase:

- The archetype of this sort of project is the LEGO kit with its step-by-step picture instructions showing how to create the object on the box brick-by-brick.
- A simple online search for "STEM kit" will yield thousands of different project kits. Subscription services will mail kits to you monthly, each resulting in a separate creation.
- There are highly sophisticated robotics kits, computer assembly kits, and furniture assembly kits that are also imitation projects, even though the finished product looks "innovative" in a way that the LEGO car might not.

There are many ways in which these kits can be deficient, though, usually in the clarity (or translation) of their instructions, but in general these kits have everything students need to do the project.

As long as publishing has existed, there have been books and magazines that shared building plans and blueprints for various projects, but the internet has exploded this practice.

- The number of YouTube videos dedicated to walking viewers through projects step by step is seemingly boundless.
- The website Instructables.com is based on users sharing project instructions for free online. The default structure of each project includes materials (often with links), tools, and instructions that often have videos or photos of each step. In chapter 6, I discuss evaluating these projects in the makerspace class.
- Several enormous 3D model repositories have millions of downloadable, user-generated objects to print, laser cut, or fabricate with some other digitally controlled tool.

As with any user-generated content, the quality of many of these shared projects can be questionable.. It can lead to frustration and disappointment when the result does not match the student's goal. There is nothing quite like browsing Instructables.com to emphasize that creating amazing things and communicating how to make amazing things are two completely different skill sets.

Digital fabrication projects with downloadable parts and instructions allow students to use 3D printers or other tools to turn the digital models into physical parts and then proceed with the assembly. The beauty of this type of project is that students learn that components can be printed or fabricated on demand when they have access to a reasonably well-stocked makerspace. Makers can take agency over materials and start seeing themselves as a part of the chain of production. On the other hand, the results from desktop 3D printers can vary, and prints that need precision may not fit together as well as traditionally manufactured parts. As with Instructables projects, the content on sites like Thingiverse.com is user generated, with all the problems and imperfections that this can cause in both part design and instructions.

Modification Projects

Modification involves students making choices about the result, so it is less common to find kit projects that involve much modification, though a good kit will leave room for modification in some way. Anyone who has built a LEGO kit from the instructions knows it's relatively easy to add to or alter the project in the instructions, as long as you have a few extra parts on hand. The essence of modification is to take something that already exists and improve or personalize it in some way.

Common modification projects include:

- Remix projects, which take existing objects or 3D models and allow students to redesign them.
- IKEA hacking, the process of upgrading or improving the Swedish furniture, is another popular modification project. You can find remix IKEA projects on 3D model websites, and hundreds of thousands of YouTube videos of IKEA hacks.
- There are many ways to personalize objects in the makerspace, from laser or CNC engraving to vinyl stickers to spray paint with stencils. Students also often modify their own creations, taking the result of an imitation project and modifying it by continuing beyond the tutorial. This experience can be an excellent way for students to learn how to go beyond the imitation skills they have gained and push off in new directions.
- Reverse engineering projects are also modification projects, allowing students to find new ways to construct an existing object. They can use 3D scanners to digitize the physical world to allow for manipulation and modification.
- Repair projects are modification projects, requiring students to take something broken and see if they can make it work again. Replacing a part, even if it is an exact duplicate of the original broken part, involves looking at the object in its current state and figuring out how to bring it back to life.
- These types of projects begin the process of students taking ownership of the world around them.

Innovation Projects

Students with experience in the makerspace and who have built a toolbox of skills and knowledge are ready to tackle innovation projects. These projects come from more open-ended prompts and design challenges that allow students to make creative decisions based on their current skill level. Even if a student poses a challenge to themselves, they should identify the goal before the steps to reach it. This is the imitation process in reverse, building backward from the result rather than working step-by-step toward a finished product.

Innovation challenges range from students free building with LEGO bricks, rather than following kit instructions, to students building and coding a robot from spare parts. Art students who want to incorporate creative technology in their work often have a clear vision of what they want and will gladly learn the skills to make their concept a reality. Many schools offer a capstone program for upper-level students, which may result in a long-term project requiring significant skill and knowledge acquisition.

Innovation projects don't assume that students already have the necessary skills to complete the challenge. Experimenting is an essential factor in an innovation project. Students should have a foundation of skills and tools they are familiar with, which will help them gauge the feasibility of their end goal. They will inevitably have to go out and collect more skills through research, experimentation, or both.

Recently, in my advanced making class, the list of projects included:

- A collapsible, modular climbing wall
- 3D-printed puzzles that resulted in tilt-ball games
- A device to record and display the acceleration of a snowboard
- An electronic target game for hockey practice
- A tennis ball–launching crossbow
- A 3D-printable extending and retracting lightsaber blade
- A custom golf bag cart

Feasibility is another critical aspect of a good innovation project. When students envision teleportation devices or faster-than-light ships, they engage in fantasy, not innovation. Imagination is important, and someday these ideas may become feasible, but an innovation that is impossible to make is still science fiction, not yet science fact. Not all technical challenges need to be solved in the innovation project, but students should be able to make enough progress toward their goal to show the project could become a reality. If students have to say, "First, we will invent time travel," they are not being innovative.

Teaching making with these three modes presents an opportunity for teachers to move from a more structured (imitation) to a less structured (innovation) process, which will be more comfortable for the new maker and the maker educator. Students with experience in making may come in with ideas and innovations they want to pursue out of the gate. Offering the flexibility for them to start with imitation projects that help them build the skills they need to work towards those goals will keep them engaged even as the teacher slows them down a little to practice necessary skills through imitation and modification.

Beyond the Makerspace

THE THREE MODES IN COOKING

When I talk to most people about these concepts, I use cooking as a metaphor. Everyone has some experience with cooking, whether their own experience or vicarious experience watching a family member or cooking and baking shows. Cooking plays a more prominent role in movies and TV shows than art making, and even commercials give us images of cooking. It might be the most universal art form in the world.

So what does imitation look like in cooking? This is the level of cooking that most of us experience. We find a recipe in a cookbook, gather the ingredients and cooking implements we need, and follow the directions as closely as possible to make the dish. An entire industry exists around boxed ingredients packaged with high-quality step-by-step instructions with pictures and descriptions.

There are both simple and complicated recipes. Following the instructions on the back of the mac and cheese box and a Julia Child recipe are both imitation projects. They offer practice building foundational skills, basic knowledge, and hands-on experience by following a set of steps. The difference between a well-made recipe from *Mastering the Art of French Cooking* and a well-prepared box of Kraft Mac & Cheese is one of degree (admittedly a large one!), not kind. This difference highlights that imitation can be a complex and rich activity all its own.

After we have cooked from instructions for a while and become familiar with various recipes, ingredients, and methods, we can start to stray from the formula in a multitude of ways:

- We may be missing a spice and come up with a substitution rather than go to the store.
- We may want to make it gluten-free or vegetarian.
- We may want to experiment with the seasonings.

This is the modification form of cooking. In modification, we are working with something that already exists and changing it to better fit our needs or preferences. When we reach this point, we start to feel true ownership of the kitchen.

There is a spectrum to the complexity of a modification project, from adding or substituting a single ingredient to changing the entire process of making the dish. To again use Julia Child as an example, she became famous for taking traditional French dishes that chefs and families had handed down for generations and systematizing them in a way that made them accessible to an entirely new audience. She did not modify the ingredients of the dishes, just the process of making them.

I had an opportunity to experience this as a child. My father was a research scientist and worked with several scientists from other countries, particularly China and India. One of his Chinese colleagues, Dr. Hu, would occasionally visit our house and cook for us. He made delicious stir-fry dishes but could not give us the recipe — not because it was a closely guarded secret, but because there was no recipe. His imitation process was to learn directly from family members in the kitchen, not by following a set of instructions written down on paper. One day my mom watched him cook for us and wrote down everything he did, estimating the amounts of various ingredients and writing out the steps in a Western-style recipe.

The odds are that Dr. Hu would never make that dish precisely the same way twice. He was continually adjusting the components based on taste rather than exact measurements. My mother modified the process of creating that iteration of the dish, allowing her, and later me, to imitate him in a way that worked for us. Over the years, this recipe has become a modification experience for me. I no longer need the recipe in front of me, and I have swapped out ingredients, increased the seasonings at times, and even substituted when I was missing something. I now feel comfortable modifying it with what I have on hand and adjusting the seasonings by taste rather than by measurement. Imitation and modification are entwined in ways that support both modes of making.

We are ready to start innovating when we have mastered a few techniques and have a catalog of ingredients and tools. An innovation project evolves out of knowledge and skills gained and a desire to put them into practice to make something new.

Innovation is never a linear process. It is more complex than adding tool A to skill B equals innovation C. It involves experimentation, iteration, risk-taking, and willingness to fail. I use the words "based on" in my definition because an innovation project should arise from a foundation of experience but go beyond this basis as ideas, questions, and experiments present themselves. When we watch experienced chefs on TV assemble some new dish from random ingredients they have been presented with, we are watching innovation in action.

While cooking is a great way to explain these modes of making to many, there are other models for these concepts built into a typical school experience that may resonate better with students: extracurriculars such as art, drama, music, and athletics.

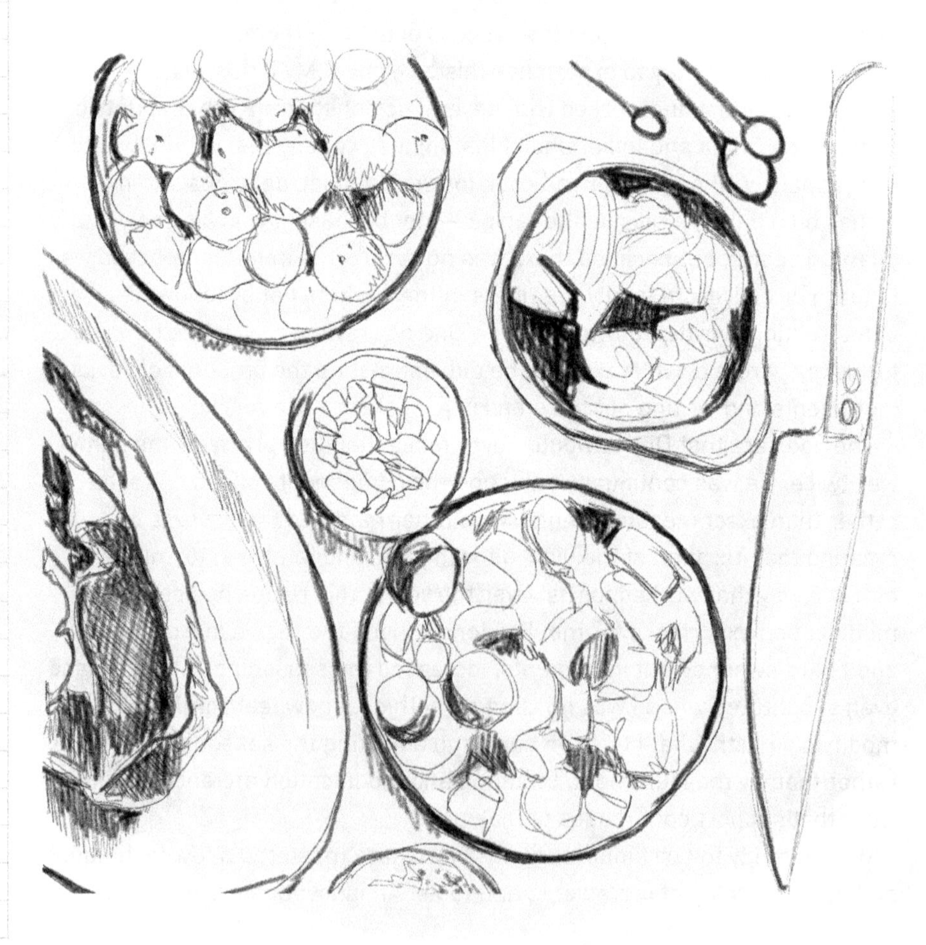

THE THREE MODES IN EXTRACURRICULARS

Students sometimes hear the word *learning* and see it in the transactional way, despite the efforts of educators to redefine the teacher/student relationship. They think, "If I do all the homework and take all the notes, I will pass the test and will have *learned* the material." In the context of making, "learning" more closely means practicing skills to gain a level of mastery. Mastery shouldn't be considered an endpoint in a process any more than a professional athlete would feel like they no longer needed to practice. Henri wrote about the idea of mastery:

An art student must be a master from the beginning; that is, he must be a master of such as he has. By being now master of such as he has there is promise that he will be master in the future.

Students who participate in athletics and the arts already understand how practicing a skill leads to ownership of those skills. Only through the practice of skills do athletes internalize and master the various elements of the sport. They know what it takes to apply them in high-stakes situations like a game, tournament, or performance.

- Imitation is learning the skill through drills.
- Modification is using those skills in an actual situation.
- Innovation is when the team leader becomes the "coach on the field" and starts directing strategy in real-time.

Many traditional educators, parents, and students do not see this process as "education." Those locked into the idea of education as teachers handing over knowledge in return for gradable performance will not see intentional teaching in these pursuits. However, all we need to do is look at the improvement in basic skills, decision-making, and leadership from the rookie athlete to the veteran to confidently say that learning has happened, no matter the method of instruction.

THE THREE MODES IN CONTENT CLASSES

I mentioned previously that many maker education programs turn into robotics and engineering programs in high school. When this occurs, the makerspace becomes a place for technical classes and leaves behind the creative practice skills that are so important to the future of maker education. This happens because of a lack of curriculum in high school maker classes, as well as how the makerspace is seen by students, teachers, and administrators.

When I was setting up my first school makerspace, I visited several local schools to learn about their programs. At each of these schools I asked how the English teacher or the history teacher scheduled time in the makerspace to do a project with their students. None of the schools had a system in place to allow this to happen. Their makerspaces were dedicated to robotics, engineering, and coding.

For years I have been a proponent of comparing a makerspace to a library, a shared community resource that supports all areas of the school to teach and learn all subjects better. A student in a high school Spanish class would not think twice about going to the library to research a topic, but they might never consider going to the makerspace to do a project about Spanish vocabulary or the culture of Mexico. A makerspace teacher that is not able to work across the school curriculum, usually because of a full schedule, is a missed opportunity for a school to enrich the learning of all their students, rather than just the few who find themselves in robotics, engineering, or even a maker class.

I have been blessed at two schools to be able to describe myself as the "librarian" of the makerspace. I teach dedicated maker courses, but much of my time is spent working with content teachers and their students to bring making into their classes, often replacing a test, paper, or presentation. These schools saw the value of promoting maker education throughout the school in various subjects and created a schedule that facilitated that work. I use the three modes of making as I design and implement projects for traditional content classes.

Imitation projects in content classes are most common in science and math classes, where following a process to arrive at an expected result is built into the curriculum. These projects might involve following instructions to create a device that demonstrates a scientific concept. In these projects, students see a hands-on application of an otherwise abstract concept. Turning theory into practice is more engaging and helps students grasp concepts in a tangible way that may stick with them better than theory and diagrams alone.

Modification projects will work across all content areas. In social studies, arts, English, and other humanities classes, where students are routinely asked to present their interpretation of material, a project where students get to personalize the product to express their viewpoint can be effective and powerful. "Voice" and "choice" have reached buzzword status in education, but they reflect the buy-in students gain when they get to personalize the content within a hands-on project. When what you make represents you and your interpretation or understanding of a topic, you are more likely to put thought and effort into the result. This feeling of ownership and pride is at the heart of modification projects.

Innovation is also possible with content-based projects, though it often takes longer to do well. In well-designed project-based learning, an innovation project involves students beginning with only some of the answers. They start with some understanding of the fundamental issues or concepts involved and then process information, do additional research, and develop their thinking on the content through the creation of the project.

A content-based innovation project can be an exhilarating, if somewhat uncomfortable, experience for many teachers since the process evolves over time and the result is not set from the beginning. Seeing the end product of engaged students thinking, discussing, and learning in a hands-on environment is a wonderful experience for a teacher. Most of the things we learn in life come from something other than a traditional classroom-like setting. Real learning comes from experimenting, failing, learning from failure, negotiating with others, discussing ideas, and all the other activities in a truly innovative experience.

Depending on the structure, imitation projects can take from one to a few class periods, a modification project should take at least a few class periods and can take as long as needed, and an innovation project may take weeks to do. With these timelines, it is helpful to think of the imitation projects as an assessment for a class unit, the innovation project as the entire unit, and the modification project as somewhere in between the two. In chapter 9, I describe a variety of content-based projects in all three modes.

Creative Process Versus Final Product

Despite the best efforts of many excellent educators, high school education is still a results-based rather than a process-based endeavor. Though there are many progressive education theories and forward-thinking schools, as of 2019, 90% of U.S. children attend public schools, most of which rely heavily on testing. This system is a culture of teachers, administrators, students, and parents. It can be harder to explain the idea of assessing a student's process, rather than the end product, to parents than to administrators.

Take, for instance, the parent who proudly told me their child "built their own computer." This task is impressive no matter how you look at it, but it can result from radically different processes. If the home-built computer was the result of assembling a kit project, connecting the monitor, motherboard, processor, and peripherals in a step-by-step process, the student has gained some understanding of the computer parts. They may not necessarily have learned anything about how some components may or may not work with others. The student who assembled a computer by researching and selecting the different parts from various sources, learning about which processor is the best (and what makes it the best) and will meet their needs, has gained a much deeper understanding of how a computer works.

It doesn't mean that the first process is not valuable while the second is; it just means that they are different, involving different skills.

- The first student has practiced following a set of steps correctly and patiently to accomplish a goal, paying attention to the details of the instructions.
- The second student has gained a deeper understanding of computer parts, probably making wrong turns along the way and fixing those mishaps with tinkering, trial and error, and research.

The second outcome might seem more desirable, but most of us will never have to assemble a computer from spare parts. All of us will have to follow a set of instructions.

The point is that each of these hypothetical projects results in a working computer, so the end result does not enlighten us about what the student learned or accomplished in assembling it. Maker education is an opportunity to shift the focus from end products to process because it still feels new, and parents have fewer expectations of what they think the teacher should be doing.

Why These Modes Matter

If we accept that shifting the focus in maker education, and education in general, towards students building skills in the process rather than the product, we might ask why it matters to identify different modes of making. Students will learn when put into an environment and given authentic and exciting challenges to tackle. In his award-winning TED talk, educator Sugata Mitra demonstrated that with access to information and resources, children can learn complex topics independently, with minimal encouragement from an adult. He calls his program a Self Organised Learning Environment, or SOLE, and has promoted it as a way to expand education into unstable and poverty-stricken environments. If children can do it independently, why focus on how it is happening?

I believe that by identifying these modes, how they work, and how we can teach them intentionally, we can level the playing field for students in these learning environments. Whether in a makerspace or a SOLE, students arrive in the space with different experiences, abilities, and inclinations. Some will easily be able to pay attention to the details of a set of instructions, while others will want to rush ahead and make assumptions. Some will start experimenting with an open mind and a growth mindset, while others are afraid of failure and hesitant to make a first step. Some already know how things work and can figure out ways to improve them, while others take things for granted and don't question the status quo. While students can learn and grow on their own, a teacher who understands the different skills, and has techniques to coach and guide students who may be struggling, is a valuable resource be an invaluable resource to those students.

Most educators now refute the idea that teachers hold the knowledge, and their job is to pass it on to their students like pearls of wisdom. Most teachers today attempt to teach their kids how to think rather than what to think. We live in a world that increasingly needs creative problem-solvers and leaders. As educators, we need to be better at intentionally teaching skills that unlock the creative potential of students, both those who already feel creative and those who doubt they can be creative at all.

3 IMITATION SKILLS

One summer, my cousins and I went to stay with our grandparents in Virginia for a week. My grandmother bought us each a paint-by-numbers kit to keep us occupied, complete with all the little plastic containers of paint colors needed to complete the image. My painting was a profile of a horse, with several shades of brown for the horse's coat and greens and blues for the background and sky. I can't remember if my cousins also had horses or if theirs were different, but I do remember that the result of my paint-by-numbers was better than theirs, enough so that we all recognized it.

For years one cousin has retold that story as evidence that she always knew I would be an artist. It has always struck me that we could get the same starting materials and procedures and end with different results. It may not have been "art," but there was some skill that I had already developed that affected the results.

I speak of imitation in the maker sense, but imitation is a standard mode in all sorts of endeavors:

- Math students have to show their work so teachers can see if they followed the steps of the process.
- Assembling flat-pack furniture requires putting the parts together in a specific order.
- Getting to a destination requires following a series of directions.

Imitation is a skill we are born with. Researchers have identified mirror neurons, which are particularly important to learning in our earliest development. These brain cells respond the same way when we see someone do something as when we do it ourselves. Mirror neurons make us feel as if we are performing the action witnessed, not simply learning vicariously in a detached way. We literally get into the mindset of the person we imitate.

Successfully following a set of instructions also requires the proper mindset. Rushing, skipping steps, making assumptions, and being distracted can lead to mistakes and frustration. When we have enough time and focus on the task, we pick up on details we might have missed otherwise. If I rush to complete the project, I may use the incorrect screw, leaving me with the wrong piece of hardware at the end of the assembly and some work to undo.

These skills are what are called metacognitive skills. Sometimes called thinking-about-thinking, these skills help set the table for all sorts of involved and complex tasks. Many talented and gifted artists and makers quit projects because they have not yet developed the metacognitive skills of positive self-talk and self-monitoring. Many exciting and interesting projects falter due to poor planning and goal setting. Many others have gone off the rails for lack of self-questioning. These are not technical skills, but they make technical skills function in the real world.

In this chapter I describe five different skills that makers practice when working in the imitation mode. They are not technical skills, though students learn plenty of technical skills along the way. These metacognitive skills are important for all projects that makers take on, but in an imitation project it is easier to see how they impact a project's success.

Finding Ideas

One of the first project ideas that a student presented to me when I started at my current school was an exoskeleton. My mind went to frames and braces for athletes or people with disabilities, but their mind was on Iron Man. Another student approached me about making the body of a car with our school's CNC machine. I thought about layers of carved foam as a fiberglass mold, but he was thinking about modeling an actual car, engine parts and all.

A little bit of knowledge can be a dangerous thing. The Dunning-Kruger effect describes a phenomenon wherein a small amount of knowledge leads to overconfidence before that confidence crashes when we realize how little we actually know. With more experience, that confidence level starts to rise again, but this time based on authentic experience, not assumptions.

The Dunning-Kruger effect can explain all sorts of incorrect assumptions adamantly held by people with very little knowledge of a topic, such as climate change and COVID-19 vaccines. We are all prone to it. How often have you read a headline or article and shared a claim with others without

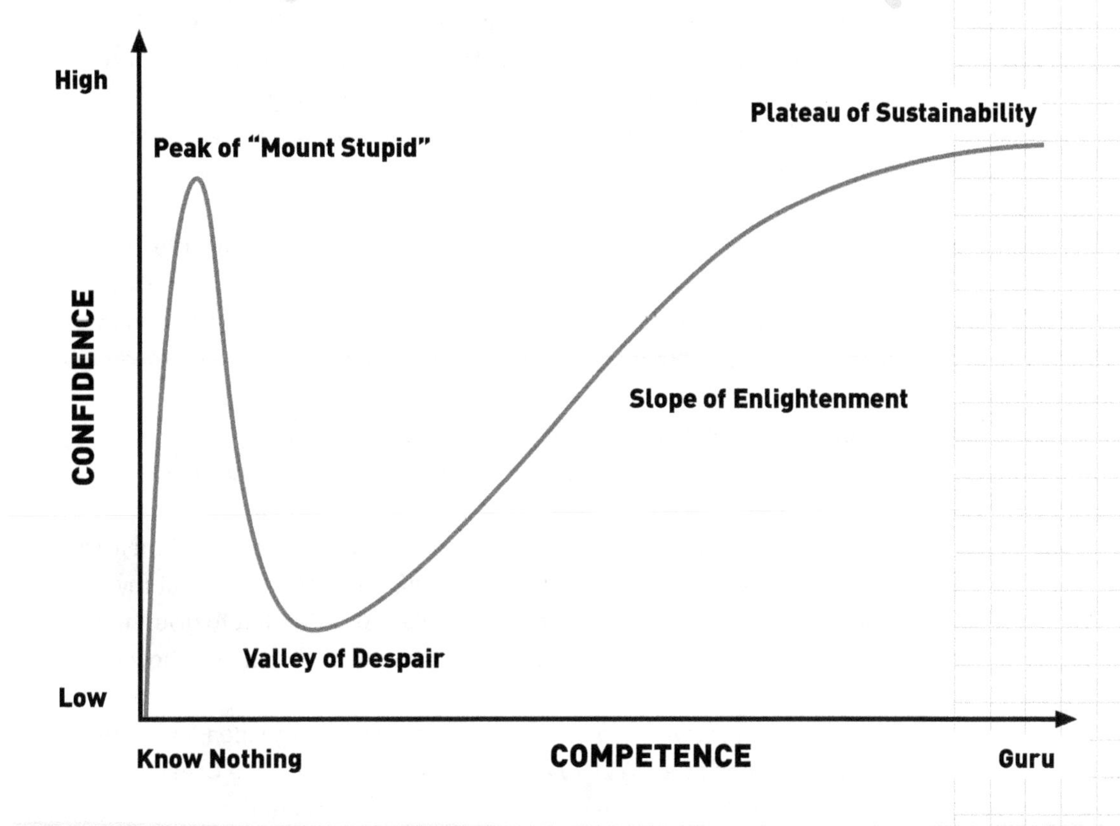

discussing alternative viewpoints or seeking additional information? In today's media-saturated world, where anyone can post "news" on social media without credentials or credibility, it can be challenging for experienced readers to filter information. For students it can be dangerous.

When a student comes to me with a project such as an Iron Man exoskeleton, it is my job to tell them the truth: such a project is not possible with the tools and time that we have. On his show *Savage Builds*, Adam Savage worked with various manufacturers to create a titanium 3D-printed Iron Man suit, and while it had none of the mechanical motors that gave Iron Man his strength, he did find a way to make it hover. The resources, experience, and budget that Savage had to create this project are vast compared to even the best school makerspace, so a student's expectations must be tempered with reality.

Students new to the maker experience have often seen amazing things created by makers with the words *3D printing*, *CNC*, or *coding* attached to them. The movie montages of Robert Downey, Jr., building his various Iron Man suits last for mere minutes. At one point, the villain in the first *Iron Man* movie, played by Jeff Bridges, tells an underling that Tony Stark built his suit in a cave with a box of scraps. When we are at the peak of "Mount Stupid," we imagine that we are all Tony Stark, not some nameless engineer working for the bad guy.

Once I have tipped them into the Valley of Despair, I do my best to start students up the Slope of Enlightenment. By encouraging them to tackle any parts of their grand idea that they can complete, they start seeing the project as a series of manageable steps. They could:

- Make the mask in plastic and give it a nice paint job
- Build the glove with LED lights and motion controls
- Work out a glowing chest piece that has a nice "breathing" effect"

Those first steps can put them on a path to gradually work toward their real goal: to make something they can be proud of. I remind myself that my job is not to help them make the next high-tech exoskeleton but to give them the skills they can build on to reach the Plateau of Sustainability, however long that may take.

For every student who comes to me with an overly large idea, I have ten that have no ideas at all. A 2010 *Newsweek* article titled "The Creativity Crisis" described a steady decline in the "Creative Quotient" of American students after 1990. The article is based on the scholarly work of Kyung Hee Kim of the College of William & Mary. The Creativity Quotient test, which psychologist E. Paul Torrance developed in the late '50s, examines a child's ability to think divergently. Divergent thinkers can develop multiple ideas to solve a problem, an essential quality in creativity.

The most significant finding of the Creativity Quotient test is its correlation to creative pursuits in adulthood. Children who score higher on the CQ test are more likely to contribute innovatively when they are older. According to the article, the children who performed best on the tests were more likely to become entrepreneurs, inventors, authors, software developers, and other types of professionals who contribute creatively to the economy. Lower CQ scores mean fewer creative individuals contributing to society, culture, and the economy.

A lot has happened since 2010, not the least of which is the growth of the maker movement and maker education. *Make:* magazine, Instructables.com, and YouTube, all huge repositories of freely available creative ideas, were all founded in 2005. The cost of many digital fabrication tools in makerspaces, such as CNC machines, 3D printers, and laser cutters, has plummeted. The first school makerspace that I created had a budget of $10,000. We bought a hobby laser cutter, a couple of off-brand 3D printers, some electronics, and got a lot of donated tools. The creative work from students that arose from this small space was staggering.

When I start imitation projects with new makerspace students, I offer them a curated assortment of projects that utilize various tools we have available in the lab. Students that struggle to come up with ideas often fall prey to "analysis paralysis," or the inability to choose between good options. When presented with the entire list of projects on Instructables.com, students will scroll for hours "considering" options. The real issue is fear of missing out, or FOMO. If they can only make one thing, what if they don't make the best one?

Students should have choices if they are to enjoy and value what they are making, so I look for projects resulting in objects that students may want. I also allow students to pick a project outside my curated list, as long as they can show that their chosen goal is measurable, achievable, and challenging. In chapter 11, I list the criteria I use to curate this list.

The best way to combat the Dunning-Kruger effect and analysis paralysis is to kickstart action as quickly as possible. Getting students engaged in a project quickly sets the tone of a bias toward action. Combining that with goal setting emphasizes "enlightened trial and error," a more intentional version of learning by doing.

Setting Goals

There are many established frameworks for goal setting, but the one I find most useful for maker projects focuses on only three characteristics of a good goal:

- Is the project **CHALLENGING**?
- Is the project **ACHIEVABLE**?
- Is the success of the project **MEASURABLE**?

These three factors should be balanced to create a good goal.

Students come to us with radically different backgrounds and levels of experience. What is challenging for the student who has done maker summer camps, has building kits at home, and happily spends their time tinkering and fiddling with things, will be very different from the student without that access. By focusing on each student challenging themselves

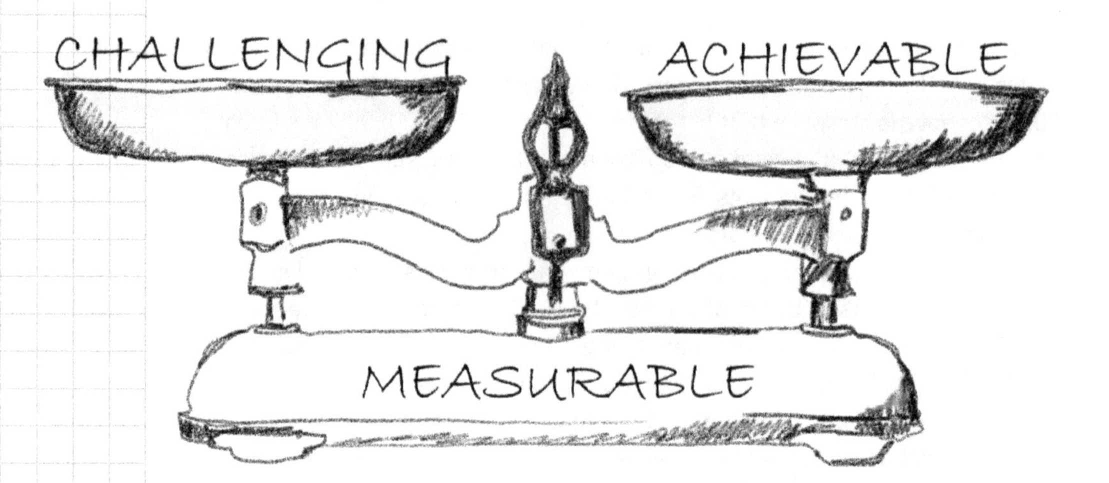

based on their ability level, we can encourage all students to push a little past where they already feel comfortable. For some students, the new skills gained might be pretty high level, while others are just learning the basics, but each student is adding to their toolbox and challenging themselves to improve and grow.

Whether a project is achievable is a determination that must be based on the time available and their current level of skill, as well as the resources the student has access to that can help meet the challenges. The student who sets out to create their computer from scratch with no previous experience, and gives themselves 10 hours to complete the project, is not giving themselves an achievable goal. Because they lack awareness of their current skill level and the project requirements, they are setting themselves up for frustration and the likelihood of quitting.

Time is always a challenge when there are hard deadlines, like a showcase that the student needs to be ready for or the end of the term. But when students (not to mention adults) don't have time restrictions, the goal becomes less concrete and may or may not come to pass. Giving yourself infinite time makes it more likely you will never complete the project at all.

The measurability of an imitation project may seem evident at first glance; did the project get completed? But there is more to it than that: This goal helps gauge a student's thinking about the overall project. One of the primary measurable features is the project's duration, which should be within the timeframe available in the course or the time set aside for that particular project. Does the student's estimate of how long the project will take come within reasonable shouting distance of the teacher's estimate? Students usually underestimate the time needed to complete a task because they assume everything will go smoothly, with no mistakes or complications. Recipes often include a time estimate, with prep time, cooking time, etc., mapped out in the directions. If it is the first time I have attempted a recipe, or the prep includes some process that I am unfamiliar with, I expect the time to go much longer than what is on the recipe.

Another measurable feature is what skill or experience the student hopes to gain through the project, or what tangible result to occur. If they are doing the imitation project to gain skills for some future innovation project, it should be clear that they have acquired those skills, not just mimicked the project steps without understanding what they are doing. If their goal is to create an object to give as a gift or put on display, the visible quality of the end product should be as high as they can accomplish.

Here is where the balancing act begins as students consider whether their chosen project is challenging, achievable, and measurable.

- A complex project may be difficult, but is it achievable within the measurable time frame?
- A simpler project may be achievable, but does it challenge the student to learn new skills and build their experience in a valuable and meaningful way?

Students must explore these considerations before, not after, starting on their project.

Setting goals is an important part of any project, not just imitation. An imitation project allows students to practice making clear goals because, with only a little research, they should be able to identify the challenges they are likely to encounter and the resources they have to overcome them.

Working Systematically

In 2011, Atul Gawande published *The Checklist Manifesto*. In this book, he proposed that complex systems need more than experts and high-quality components to eliminate mistakes; they need checklists.

We live in a complex world that is only getting more so. Humans and machines work hand-in-hand to do amazing things in various fields. Dr. Gawande is a surgeon, and his book and TED talk speak from the perspective of healthcare, where mistakes can lead to death or lifelong health challenges. Developing the habit of using checklists through low-stakes — but still complex — maker projects is a way that maker education can help prepare students for work in a variety of fields.

In maker education, students practice the skill of developing a checklist to ensure that a project progresses smoothly. Imitation projects where students make multiple copies of an object are an excellent opportunity to have students develop a checklist that helps them identify critical steps, challenge their assumptions, and avoid mistakes.

Dr. Gawande identifies two different types of checklists: READ-DO and DO-CONFIRM. In both cases, the list should be at most 5-9 items, which is the limit of our working memory:

- **A READ-DO** checklist requires team members to read the items aloud and then do them one at a time, making sure that each step has been carried out.
- **A DO-CONFIRM** checklist is for people already experienced in the task but who can benefit from double-checking to avoid mistakes.

Makers often find themselves in a groove when working on a repetitive project. We start and stop machines, drill holes in the exact location, and make jigs to simplify the measuring and assembly of objects. We often develop a rhythm and muscle memory that can carry us through a process. But one change in that rhythm can throw us off. If someone stops by for a quick chat, we might forget which step we were on. If a machine acts up and we have to fix a printer jam, or a piece of wood has a tricky knot to deal with, our easy flow gets interrupted. These moments are when we can skip a step or make a mistake. A well-designed checklist can help us get back on track.

Developing checklists is a worthwhile skill for future work and gives students experience in thinking about systems and how to make them work efficiently and effectively. In a world that continues to grow more complex, we will increasingly rely less on the lone expert and more on collaborative teams that rely on good checklists.

Learning From Mistakes

A good checklist can help us avoid mistakes in complex situations, but how do we learn those critical steps? The most likely way is by discovering an error several instructions past the crucial point. Realizing you can't proceed in a process because of a small mistake is a strong motivator to slow down and pay attention. Having to redo some portion of the project reinforces the lesson. Often these minor errors stick in the brain and are remembered long after the project is complete.

When a mistake happens, students often blame the process or the instructions rather than considering how their own actions played into it. It may be that the instructions were flawed, but why didn't the student recognize something didn't make sense? Seeing mistakes as an inevitable part of the process is important for when students take on more complex challenges.

Students almost always blame the computer when their code doesn't work. They personify the program as an adversary rather than face the fact that they have made a mistake, either in their thinking or their code structure. Even when they discover the mistake and realize that it was their error, they still find ways to blame the computer. The reality is that computers can only do what we program them to do. Eventually, when the code doesn't work, they start to hunt for errors, not blame it on the program.

Learning from mistakes and how to accept that we all make them is an essential aspect of maker education. Students may not want to make mistakes in math because they will lose points, but when they try to make something meaningful to them, they will want to overcome mistakes because the result matters.

This experience is not restricted to imitation projects, but learning to isolate the error in a step-by-step project is much easier. Students are less likely to get frustrated by the need to undo and redo work in a lower-stakes imitation project. Learning to correct mistakes as early as possible in the process, rather than avoiding dealing with it, yields a better result in the long term. Catching the mistake as it happens is the best result of all, and this takes practice.

Paying Attention to Details

The end goal of becoming skilled with imitation projects is an improved ability to pay attention to details. Focusing on technical skills and knowledge acquisition is more manageable when details become less of a stumbling block.

Simply watching a tutorial is not enough to collect and internalize all the details. To understand the process, the hands must be involved in imitating and following the steps. We have all experienced the "aha" moment when we realize we missed some step or detail and discover it was right there in the instructions. We would have seen it if we had slowed down and reduced distractions.

When fully engaging in a step-by-step project, there is a point where you can feel your attention narrow to the set of instructions in front of you. Through this narrowing of focus, we reduce mistakes and gain valuable insights into the process we are imitating, recognizing patterns and strategies we can apply in the future.

Makers and artists often describe this narrowing of focus as a state of flow. It can be quite difficult to get into this state in a typical one-hour class block, though. School schedules that use longer blocks make it easier to reach that flow state. A strategy to deal with short blocks is to lump activities, discussions, and lessons in one day, so that other days are devoted exclusively to working on projects.

For some kids, experiencing this state of flow is enough to recognize its value in their lives, whether they apply it in the classroom, the athletic field, or the makerspace. Others will need to learn active strategies to make it work, like shutting off their phone, moving to a quieter space, or selecting a project that aligns with their interests to increase their engagement.

When students have mastered slowing down, they can apply this to projects in more advanced settings, but it is hard to learn how to slow down when engaged in more complex tasks. Imitation projects offer a valuable opportunity to practice this skill so that it can be easily applied when the stakes become higher.

4
MODIFICATION SKILLS

When I was 11, my parents signed me up for a drawing class at a local visual arts center. I still have the sketchbook they bought me, which sits on a shelf in my studio next to sketchbooks from college and graduate school.

One of the first tasks was to draw two wooden blocks, one short and one tall, standing on their ends. My first attempt was sad. The lines of the sides of the blocks are curved and lean over instead of standing straight, and the perspective is wrong. They look like blocks made of Jell-O instead of wood. I can still remember the frustration that the drawing didn't look the way I pictured it in my head.

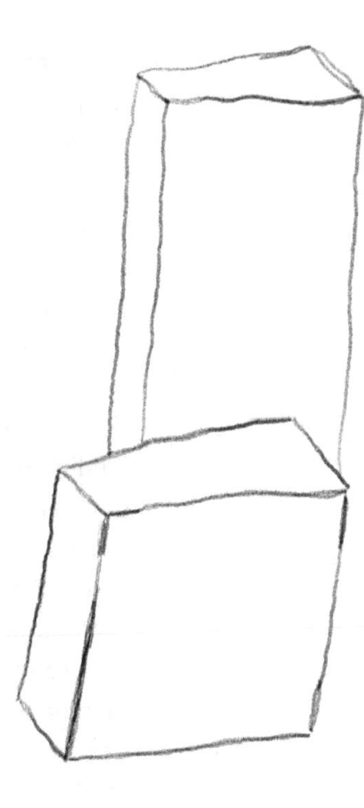

The teacher told me that if I wanted to draw something from life, I needed to rely on what my eyes saw, not what my mind thought was in front of me. In drawing, as in many other pursuits, we often make assumptions about how things are put together, ignoring the evidence of our senses. I drew the blocks again, this time making my hand follow my eyes, and got them right. The lines were straight, and the perspective and proportions were good. I was so happy with the result that I continued adding the blocks' slightly beveled edges and drew in the wood grain. When we continued the exercise and stacked up a more complicated arrangement of blocks, I could apply what I had discovered to the next drawing and the one after that.

If imitation involves metacognitive skills to help us manage a process, modification requires students to deal with the world as it is. In order to modify something in a practical or meaningful way, we have to understand its essential qualities. These qualities include physical attributes like material, size, shape, etc. There are other qualities that are less tangible that are just as important, like aesthetics, emotional connection, and meaning.

Measuring and modeling are important in modification so that our changes work with the existing design. Digital fabrication allows us to model our ideas on a computer before we ever generate a physical model, thereby saving time, resources, and money. Learning how to make precise and useful measurements of a physical object is extremely important whenever students hope to make a product. Even when working with code, students need to be able to measure time, voltage, and sequence.

Finding out a user's needs, rather than making assumptions, is a valuable aspect of modification. Developing empathy for a client or user helps make our ideas better, not just in form but in function. There are so many amazing inventions that functioned perfectly well, but never worked for actual people.

Finally, when we really understand an object, we can start to see its potential and how it might be used for more than its intended purpose. Finding new uses for existing products, materials, and designs is a skill that can be developed through practice. Bending and twisting the existing world into a new and more useful shape is a major element in invention and innovation.

Whether simply altering the look of an object or redesigning it from the ground up, students can practice many worthwhile skills through modification. These skills involve learning how to take the world as it is, gain greater understanding of it, and make it more meaningful to ourselves or others.

Personalizing and the IKEA Hack

People love to mark things as theirs. Whether signing artwork, getting monogrammed shirts, or posting an image on Instagram, we like to take something and claim it as ours. Behavioral economist Dan Ariely has studied the value we place on things we make ourselves versus similar items others make. Even when we do a poor job, we overvalue our work.

Ariely describes the earliest version of Betty Crocker cake mix, which only required the home baker to add water. Sales of the mix eventually waned because women said they didn't feel like they were baking the cake. The company changed the mix so that you had to add eggs, oil, and water and focused marketing on the decorating. These changes allowed the home baker to "make the cake their own," and the products have been strong sellers ever since.

Ariely equates this story to building IKEA furniture. When we have to put effort into a project, we feel a sense of ownership. We become more attached to what we make, even if it looks or tastes exactly like everyone else's version. He calls this "the IKEA effect," and it explains much of the appeal of the maker movement and DIY culture.

A related phenomenon, the IKEA hack, is driven by the skills of

modification. IKEA furniture is inexpensive and designed to be efficiently packed in flat boxes. The assembly instructions are almost entirely pictures so consumers can understand them worldwide without translation. IKEA offers many styles, but individual pieces have relatively few options in order to keep costs low. If you want your IKEA Kallax shelves to look different from the millions of others, you will need to do a little IKEA hacking.

The IKEA Kallax shelf is a simple set of stacked cubes in various formations. IKEA hackers have added paint, legs, doors, drawers, bins, and more to make them more attractive or functional. They have been laid on their sides, stacked on each other, and used as bookshelves, cabinets, vanities, and much more. The lengths to which some people have gone to turn this humble design into unique pieces are remarkable, and they love to share their ideas on YouTube or Instagram. College students might leave their Kallax shelves behind when they leave the dorm, but these IKEA hackers will likely always hold onto their creations.

When students find ways to become personally invested in their projects, when they make something that they want to see in the world, they push harder to overcome challenges. This motivation could be a personal desire, the desire to help someone else, or the desire to solve a problem that matters to them. Personal motivation is more sustaining than the goal to receive a good grade.

Working with Constraints

Constraints and restrictions can generate creativity. In painting, many artists use a technique called "limited palette." Instead of the entire spectrum of colors laid out before them, the artist picks only a few color options with which to work. Creating an image from limited options forces you to come up with new ways to mix and arrange them to create the effect or the picture you want. When you have unlimited options, these creative solutions don't arise. As the saying goes, necessity is the mother of invention.

In modification projects, constraints are critical. Even a personalization project as simple as a vinyl sticker on the back of a phone case requires considering measurement, color combinations, the case's material, the camera's placement, and more. Some constraints are essential to the phone's function; you shouldn't cover the camera just to put your favorite band's logo on your device. Others are personal preferences, such as the color combination of the phone and sticker. All of them should be considered.

We live in a world of immense technological invention, and it often seems that just as we think about a neat idea, it suddenly exists. Kids (and adults) who have never been a part of a long-term project have difficulty conceiving of all the work that went into developing that final product. It can seem as if smartphones, 3D printers, and new apps spring up as if by magic. The first 3D printers were created in the early '80s, and teams of engineers and designers gradually overcame the various constraints so that you could order a consumer 3D printer online for a few hundred dollars today.

As projects get more complex, the constraints become more challenging. Many students need help to focus on only one or two aspects of a project at a time. I often do the "Marshmallow Challenge" to demonstrate the importance of identifying and dealing with the most critical constraints early. The Marshmallow Challenge puts students in teams of four or five and gives them 20 pieces of dried spaghetti, a yard of tape, a yard of string, and a marshmallow. The teams get 18 minutes to build the tallest tower

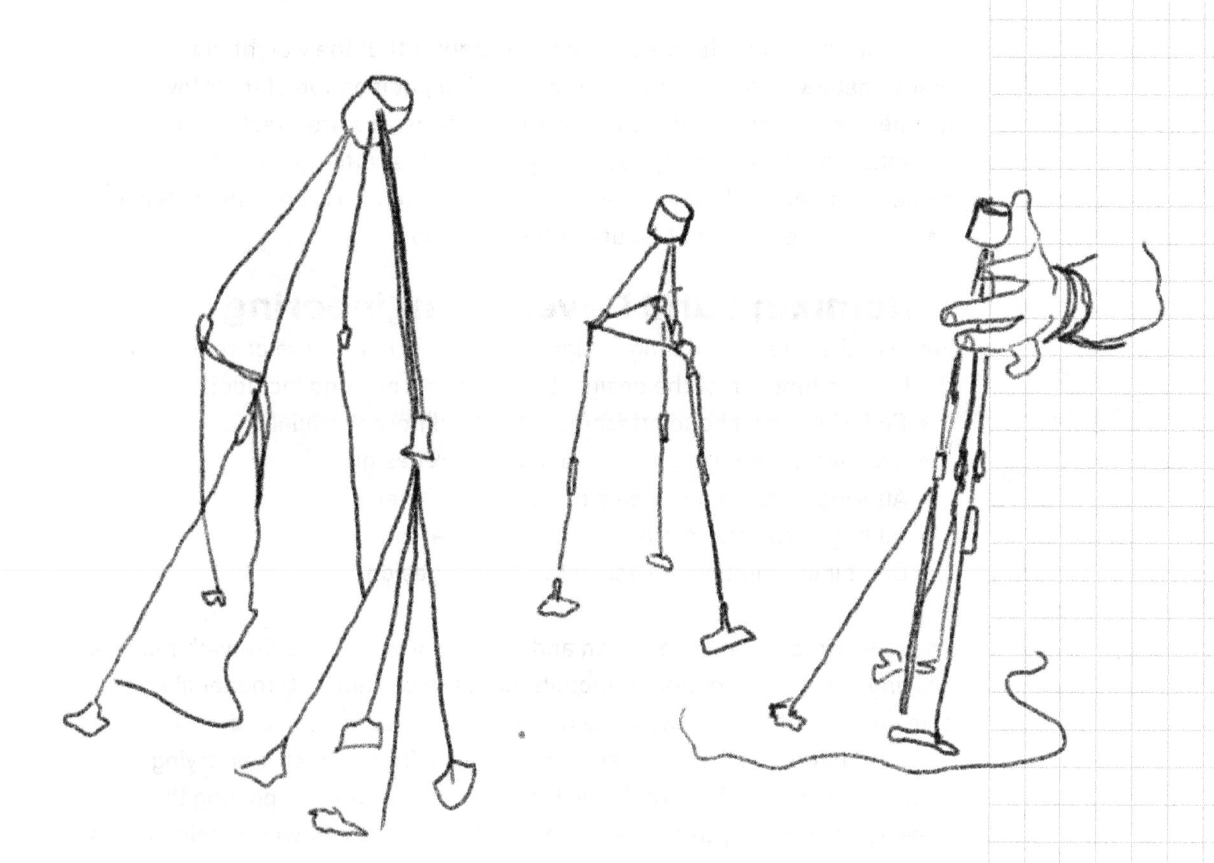

they can that is able to support the marshmallow at the top.

This is a simple challenge, but there are several constraints that play into the challenge:

- Dry spaghetti is a very brittle building material.
- The string is strong, but how do you use it effectively?
- How small should we rip our tape to save material but still be effective?
- How should we balance the height with the width to get a tall versus stable structure?

Teams that can identify the most critical constraint of the challenge — the weight of the marshmallow — are the most successful at the challenge. When they realize this is the critical element, they start putting the marshmallow on the tower earlier, testing to see if their early ideas can meet the challenge. Teams that don't recognize that the weight of the marshmallow is key wait until the very end to put it on top of their tower, just before time runs out, usually with disastrous results.. Testing the most essential constraints early, expecting that the first attempts will fail in some ways, means that you can solve those issues early on in the process and reduce the chance of failure at the deadline.

Remixing and Reverse Engineering

When makers remix existing designs, the goal is to improve or reimagine the form or function of the design. Examples of remixing include:

- Redesigning a phone attachment to fit a different model
- Creating a simplified version of a complex design
- Altering a character to be a different character
- Adding decorative elements for a personal touch
- Combining multiple objects into a single design

The maker movement is an open and collaborative community, with makers sharing ideas freely online. Although there are certainly 3D model files for sale online that are not available to remix, Thingiverse.com and Printables.com are full of thousands of files available for free. Rather than trying to squeeze every nickel out of these designs, the creators are posting their ideas for community use. Mostly what they want is credit when their work is used in certain circumstances.

In 2002 a nonprofit organization, Creative Commons, was formed to help increase the availability of openly-licensed creative works. Many makers on Thingiverse, Printables, and other maker websites use CC licenses on their posted works. These licenses often allow two key uses:

- **Share** — copy and redistribute the material in any medium or format"
- **Adapt** — remix, transform, and build upon the material

This ability to make copies of another maker's work and change it to meet your needs is a critical feature of the maker movement and maker education. Remixing 3D models is actively encouraged and recognized.

In the arts, one of the worst things you can be is *derivative*. Copying the work of teachers or historical figures is good for learning the basics, but the mature artist is expected to seek out originality. In schools where the art class is the only means for learning creativity, many students feel unable to be "original," and the result is they believe they are not creative.

In a makerspace, and especially in a remix project, any student can develop a clever improvement on an existing idea. Even small changes to a design can make them more functional, beautiful, or desirable as objects. Printables, Thingiverse, and other 3D model sites link remixes to the previous design, demonstrating a chain of creative modifications. Students who experience this start to see themselves as innovative in a way they may never have considered before.

A companion process to remixing is reverse engineering. This is the process of taking something apart in order to understand how it functions or how it was made. This process can get a bad rap as it can be used for intellectual property theft, but as a learning experience, it can be valuable.

The ability to 3D scan physical objects, whether through apps on a smartphone or dedicated devices, is something that many people don't realize exists. In general, the quality of these scans is low, but there are high-quality methods out there. The Smithsonian Institution has digitized many of their artifacts, some of them quite complex. In 2019, Adam Savage did a live build at the Air and Space Museum that recreated the hatch of the Apollo 11 command module. The parts were made by over 40 makers around the country from CAD designs created by Andrew Barth, an intern at the museum. In a live event, Savage assembled the components into a life-size replica, including one part that I made with a group of summer camp students. Even though the parts came from dozens of different makers across the country and were based on digital models and diagrams, the parts all fit together because of the precise measurements, scans, and digital models that Barth created.

Reverse engineering gives students practice in breaking apart complex objects to understand their properties and determine how they can be fixed, improved, or copied. It turns a "product" into a group of components, each of which can be understood.

My makerspace is located in a new building with custom signage at every door. Some of these signs have the names of the teachers who work in those offices embossed on them. In one case, where the teacher on the sign had left, the new teacher reached out to my students to recreate the sign with her name instead. The company that made the signs no longer existed, so purchasing a new one wasn't an option. Through many versions and methods, they were able to make a pretty close replica using the laser engraver, acrylic, and wood. We are now able to fabricate as many replacement signs as needed because these students used reverse engineering to understand an existing object.

Human-Centered Design

It's actually easy to come up with solutions for problems. Most of us say, "You know what they should do ..." regularly. Access to the right tools makes these solutions possible for those motivated enough to make them happen. However, the solutions we create are often designed to work for our own lives. It's harder to create solutions that work for other people.

A cognitive bias that plays a role in the makerspace is False-Consensus Bias, the belief that others' thoughts, feelings, and judgments are similar to our own. In other words, if I see a problem or solution to a problem, I assume most other people probably see it the same way. This tendency leads to many preconceptions about how to design things on the part of students, not to mention adults, and ignores that most of the world lives very differently than we do.

False-Consensus Bias

The tendency to assume that one's own opinions, beliefs, attributes, or behaviors are more widely shared than is actually the case. A robustly demonstrated phenomenon, the false-consensus bias is often attributed to a desire to view one's thoughts and actions as appropriate, normal, and correct.

It's hard to challenge our assumptions about the world and the people in it. It is easy to think that an entire group, whether political, racial, ethnic, or even all of humanity, acts and feels like we do. It is only through making efforts to understand the world as it is, not as we assume it to be, that we can solve problems in ways that work for actual people.

Anyone who has assembled a piece of hardware has experienced either good or bad human-centered design. Sometimes, the instructions are so clear and seamless that the process is almost joyful. With their massive design teams and budgets, electronics companies have redesigned the experience of setting up a new device so that you think of the entire experience positively. A more budget-friendly option, or a device from a less competitive sector, may not be so pleasant. These companies likely don't have the resources to engage in human-centered design, relying instead on the engineers and designers who built the device. Instructions written by engineers can be notoriously bad because they assume the consumer will understand what seems simple and obvious to them.

When I work with students on human-centered design skills, we talk a lot about interviewing. Interviewing someone to try and understand a problem from their perspective is an involved process that requires understanding how to ask open-ended questions and active listening. There are activities for teaching interviewing skills in chapter 11.

When students engage in human-centered design, they learn to challenge some of their preconceived notions and solutions based on false-consensus bias. When they realize that even their classmates, who likely live similar lives to their own, have very different perceptions and experiences, they start to understand that someone across the country or around the world may require a very different solution to a problem.

In addition to interviewing, empathy studies and user-generated feedback are important to see how our ideas work for people. Getting the person we are designing for involved in the process in early prototypes is a crucial step in ensuring that we aren't getting lost in the echo chamber of our own ideas. If you wait until your design is complete, it may be hard to backtrack if the student made significant assumptions about what their client really needed. User generated feedback could take the form of:

- Surveys
- Critiques
- Focus groups
- Social media posts
- Observations
- Reviews

Empathy studies involve putting yourself in the shoes of the person you are designing for. There are many design challenges for students that involve designing assistive devices, innovations around the United Nations Sustainable Development Goals (SDGs), or solving local challenges in your community. All of these require some level of empathy study to try and understand the issue from the user's point of view. If you are designing something for a blind person, you should spend time attempting tasks while blindfolded. If you are designing for someone in a wheelchair, try spending time in one. Your tired arms will give you valuable feedback about how heavy your design should be.

Human-centered design is a challenging but very valuable experience for students. Students who learn to question their assumptions about people, and to try and understand problems from others' perspectives, may become not only be better problem solvers, but also better people.

Upcycling and Do-It-Yourself

We live in a world full of objects we are not supposed to understand. Today's technology, such as phones, laptops, and cars, are designed not to be fixed by the average consumer. Most people would never consider opening up their phone to replace a defective component.

When I was in high school, a computer was still a big box with screws that could, and often did, get opened by the curious. There was space in there to add things and modify the configuration. Laptops today are so thin that every space has been designed for a specific model of each component.

Yet, the do-it-yourself enthusiast has never had as many options at their fingertips to upcycle objects. Components for various projects can be shipped worldwide in just a few days. Building projects with wood and hardware of every sort can be accomplished with a short trip to your local Lowe's or Home Depot, and there are YouTube videos with tips and tricks on how to work with them.

A person who sees the world as something that can be modified finds opportunity in the objects and materials around them. I had a roommate who went to the local dump's drop-off station to pick through furniture left by departing college students and others leaving town after the semester. Over the years, he collected a nice set that he sanded and refinished, bringing them back to life almost for free.

As we've seen with the IKEA effect, we value objects more when we have a hand in their making. If this is true of simply following the instructions to assemble a piece of furniture, imagine how much more value we place on something we have revitalized and kept out of the landfill. Students who learn modification skills can look at objects and see their hidden potential.

Another example of upcycling that often gets overlooked is redefining the purpose of an object. This skill, like divergent thinking, is another one that young children have and that they slowly lose over time. Anyone who remembers making a pillow fort out of couch cushions or turning a large cardboard box into different playthings has already done this. As we grow older, we classify items based on marketing and media. It takes a special mindset to see something in a new light.

When students learn they can alter and modify seemingly unchangable materials, they begin to see that they can be more than just consumers. Broken objects start to look like things that they can fix, house projects become things they can do themselves, and they take pride in the things they didn't have to buy. I've had students with broken backpacks, belts, sports equipment, and various other items stop by the makerspace to learn how to fix them. For students who have practiced modification, these ideas bubble up constantly and can lead to big projects — sometimes truly innovative ones.

5 INNOVATION SKILLS

In my childhood sketchbook, there are pages of drawings of faces from the front and in profile. The proportions of the human face are difficult for young artists to get right, and there are tricks to help place the eyes, nose, and mouth in the proper locations. These proportions were something that I had yet to learn, so the pages are visual evidence of trying to work them out for myself. I remember sensing that my drawings didn't look right at the time, and I wanted to figure out the issue and how to fix it.

Children drawing faces for the first time tend to locate the eyes too high on the head. We look at facial expressions, not foreheads, so we overemphasize those parts in drawings. In my sketchbook, the eyes gradually start to move lower and lower on the head, and the heads become rounder and more egg-shaped rather than having flat skulls. At the time, I felt like I was solving big artistic problems, even though the proportions of the head have been formalized in Western art since the Renaissance. It wasn't until years later that an art teacher showed me a procedure to locate the eyes, nose, and mouth by dividing the head into specific

Initial attempts with incorrect proportions

Later drawings with properly placed features

proportions. It didn't matter that this was well trodden artistic territory. Through the process of exploring how to draw faces, I experimented, failed, tried again, and eventually learned something new. This is the essence of innovation.

Innovation is a challenging word to define. A key component is creating something new, either an object or a process. Often, people assume that innovation is high-tech, but that is only one possible outcome. The first iPhone was an innovation. It was significantly new and different from the cell phones and iPods from which it arose. Every other iPhone since has been a modification, not an innovation. Incremental improvements in technology and design have made each new iteration better, but there is nothing significantly new about them.

Teaching students to be innovative is the stated goal of many school technology and maker programs, but unlike imitation, there is no straightforward recipe to do this. Innovation is something that emerges when the conditions are right.

A makerspace will have dozens or hundreds of tools. These will range from the low-tech (scissors, hammers, screwdrivers) to the high-tech (3D printers, power tools, laser cutters). Some will require training to use safely or well, and others won't. That same makerspace will have a variety of materials, again ranging from low-tech items such as popsicle sticks, glue, and nails, to more complex plastics, metals, resins, and more. Students must have gained familiarity with at least a few of these through imitation and modification projects in order to start innovating.

They will also need experience in skills they can draw upon, both technical and creative process skills. Although innovation is not an endpoint or the top of the ladder, as we sometimes assume, it does depend on experience from the other modes of making. Students need a basic understanding of makerspace resources, foundational skills, and a sense of how the world can be manipulated and improved to start thinking innovatively.

Being able to create something new is only part of the process, and there are many skills students will develop even as they are working on innovative ideas. These skills are the goal of the student who progresses through a well-designed maker curriculum.

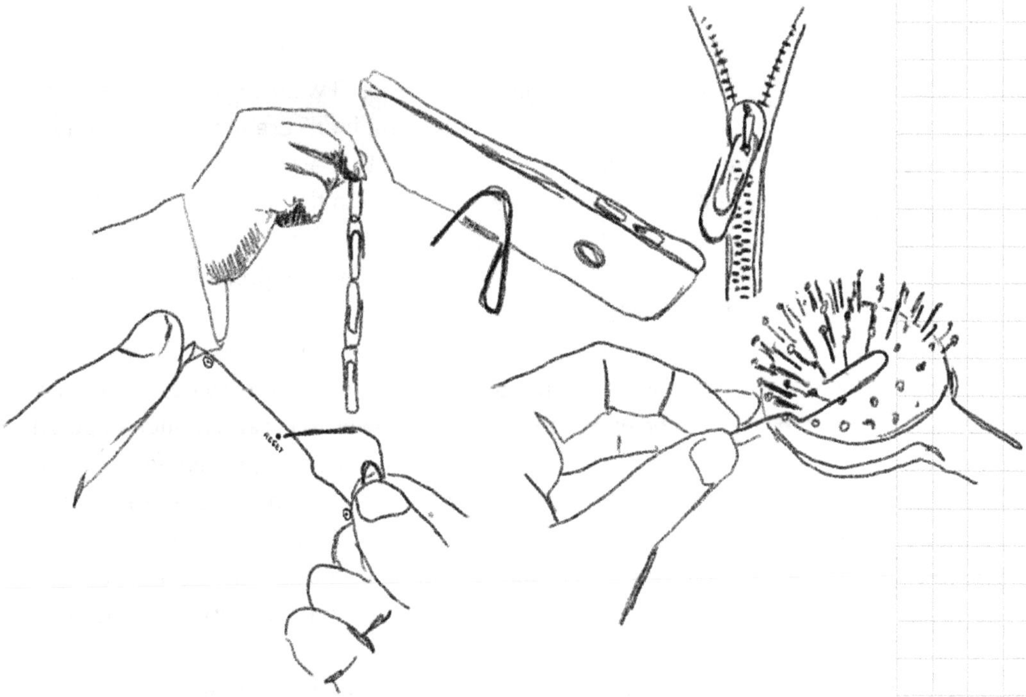

Generating Ideas

In 2011, Sir Ken Robinson conducted a study on divergent thinking for the Royal Society of Medicine. His simple question: How many uses can you think of for a paper clip? Robinson found that while most people came up with 10-15 uses, some named as many as 200. Young kids generally could come up with a lot more than high school students or adults, which led Robinson to conclude that traditional education reduces creativity rather than promotes it.

A paperclip is a few inches of metal wire folded in three places. How could you possibly find 200 things to do with something so simple? The paperclip study is all about generating ideas.

As opposed to finding ideas, which was discussed in chapter 3, innovation projects require students to generate them. This is where big ideas and divergent thinking can be an asset. Hopefully, by this time, students understand what is possible with the tools and skills they have developed. This doesn't just mean technical skills, but maker skills such as goal setting, attention to details, and working within constraints. Students who have practiced this creative process are ready to gain new skills to meet the needs of their vision.

Often this process begins with "what if" or "I wish": I wish I had X, or that X existed. "What if I used" or "what if I combined" are also great prompts for innovation. Identifying the things in life that are interesting and fun for students will help them be more engaged with the project and starts them with some concrete experiences upon which to build their idea. Kids who love sports can come up with great athletic-related projects. Kids with younger siblings might have great toy ideas.

Innovation projects don't have to look "innovative." They don't need to be high-tech or involve AI, though they certainly can. They need to be new, and they need to be within the student's power to create, though hopefully not without some significant stretching. NASA engineer Lonnie Johnson has worked on some of the highest-tech projects imaginable — and he also invented and patented the Super Soaker water gun. The Super Soaker is an excellent example of "what if": "What if I had a water gun with a super strong stream?" His inspiration came from his work in other, more "innovative" areas.

Another way to push a project from modification or imitation to innovation is to use the improv comedy method of "yes, and." In this exercise, two participants keep adding to, never subtracting from, what the other says. The idea is to keep improving the scene by being open to what was just said and then adding something to keep things moving. In an innovation project, this may take the form of starting with a simple or existing design and saying, "yes, and" to add elements until the project begins to feel like something new. Here are some examples of this from student projects:

- I would like to make something for my putter. *Yes, and* I want to use lasers to help me aim. *Yes, and* I want it to attach and detach from my putter with magnets.

- I would like my feet to stay warm in my boots. *Yes, and* I want a heater in the sole of the boots. *Yes, and* I want it to be rechargeable.

- I want a light for my room. *Yes, and* I want the light to change colors. *Yes, and* it should change colors when I turn the light to different sides.

- I want something that can make my lacrosse stick look cooler. *Yes, and* I should be able to add fun decorations to it. *Yes, and* I should be able to attach them before the game and detach them when it's game time.

- I want to make a pattern of LEDs. *Yes, and* I want it to change based on sound and volume. *Yes, and* I want it to be built into a base for my smart speaker that looks like a gramophone.

All of these innovations start with a relatively straightforward idea and grow through the process of adding elements that push the project's boundaries. One push might constitute a modification project, but once you have added a level or two of "yes, and," the idea starts to become something new and complex.

Taking Calculated Risks

Traditional education is not a place to learn how to take risks. The idea of standardized tests and meeting benchmarks is based on giving kids the same experience and comparing the outcomes. Even when students "go beyond" the expectations, they do so with activities laid out beforehand. Taking risks is often more akin to a difficult imitation project or perhaps a modification project.

There is a lot you can read about grit and resilience in kids and about fostering a growth mindset, but the basic structure of content delivery in most schools does not support this. Maker education presents an opportunity to embed experiences that help kids build these essential skills. When students first enter my class, they ask me about how they are graded or when things are due. When they reach the end of the course, our conversations are about how they can get their projects done.

Taking risks in the makerspace is not about being reckless, but it is about trying things that are fairly likely to fail. The design firm IDEO promotes the idea of "failing often to succeed sooner." When projects become truly complex, as opposed to just complicated, even an expert can't hold all the variables in their head. An innovation project will have many false starts and failed experiments. The data gathered from these calculated risks will increase the likelihood that the overall project will be a success.

High school students need to have the space and freedom, even the push, to try something that might fail or be incomplete. In an innovation project, they often work on things the teacher has never seen or done before. This situation can be uncomfortable for them and the teacher. It's important to constantly remind them (and yourself) that the goal is learning the skills of the process, not the final result. A student who creates a wonderful object but doesn't push themselves to try something new and risky is missing out on the true benefit of the experience.

The teacher's job is to keep reminding students to use the skills they have practiced. If they need to learn a new tool to create part of their innovation project, suggest they do a quick imitation project to learn how it works. Identify the constraints involved in the project and have them create a checklist to help them master a complex process. When they put these skills into practice in an innovation project, they will start to feel that taking risks is simply an important part of the innovation process, even when the attempts fail.

One of the most important things to communicate to students when trying difficult things is that the end product is not the expectation of the course. Focusing on the end result would be like saying to a young student-athlete that only a championship is an acceptable goal for a successful season. Many great athletes spend years on losing teams and still have a chance to excel at the highest levels of their sport. Those losing years must be spent practicing technical skills as well as the "soft skills" of the sport: leadership, patience, awareness, etc.

When maker educators build rubrics that define what the finished product must include to be "successful," we highlight the end product over the skills practiced to get there. We are, in effect, falling into the standardized exam model of education. I would much rather see a student set a challenging goal, practice all their maker skills to pursue it, and have an ambitious but unsuccessful or incomplete project. Although the final result may not be what they envisioned, they will have learned valuable skills along the way — skills that may make the next project a success.

Prototyping and Iteration

Teachers should build failed attempts into the structure of any project, which means that the project's timeline and production must revolve around prototypes and iterations. If a student can solve a difficult project in one version, we are probably looking at an imitation project, not an innovation project. As stated before, an innovation project is not about how high-tech or sophisticated the final result may be; it is defined by the skills used to arrive at the finished or unfinished product. Building a house from a set of blueprints is difficult, but it is not an innovation project.

The best way to wrap your mind around a complex problem is to wrap your hands around it first. This means making physical representations, or prototypes, that get progressively closer to the end product as they go along. Students should make the initial prototype as quickly as possible with readily available materials.

A wonderful video from IDEO shows a rapid prototype for an Elmo dancing app for kids. The video shows a person on a phone screen, and a hand enters the frame and taps the screen, and the character starts to do a simple dance while music plays. The hand enters again, taps the screen, and the dance changes while a narrator describes the idea. After several repetitions of this, we get to see the secret. The "phone" is a large printout of a phone with the screen area cut out. An actual person stands behind and is seen through the cutout. Writing the code for an app mockup with a moving character might take a long time. Cutting a hole in a piece of poster board and printing an image takes a matter of minutes. Both would give a visual of how the app will work, but the IDEO video shows how a little creativity and resourcefulness can save a lot of time.

Cardboard is an ideal material for early prototypes. When you make a prototype out of cardboard, you can start to think about size constraints and quickly test out many different ideas before you even sit down with a CAD program or machine tools. A prototype that doesn't quite work in cardboard can be remade and tested many times before a 3D printer completes its job.

As an art teacher, I generally designed projects that took two to three weeks to complete. It became a pattern, and eventually, it became a bit of a rut. To change this, I set up a still life and told my students that they would do a painting a day for three days. At first, they were shocked at the idea of making a painting in a single class period, but once they started, they expressed how much fun they had in just getting to work. I told them they would not be graded on traditional art metrics like perspective, line quality, correct color, etc. I was looking for them to give it a shot. The result was some of the best painting they had done all year.

Each project I mentioned earlier in the chapter, from the heated boots to the laser-guided putter, started with a cardboard prototype made in one to two class periods. The students gained valuable information from making these rapid prototypes. The low stakes of cardboard helped them move past the "analysis paralysis" that many kids experience when starting something new. It sets the bar at "making progress towards your goal" rather than "designing something ground-breaking."

After this initial prototype is complete, students need to evaluate it. Prototypes don't exist for their own sake; they exist so we can learn something to help us continue that progress. We might learn something about the form of the design, the materials needed, the next steps to take, or any number of other helpful bits of information to move the project forward. When modifying an existing idea or object, the next step can become pretty clear. When students are innovating, they are figuring things out as they go.

This early stage is a good time to remind them of the Marshmallow Challenge described in chapter 4, and perhaps to even do it again. Solve the most significant problems early on. Identify the most critical constraints and start tackling them. If a project involves coding an Arduino, and this is something new to them, they should not save this for the end of the project. That creates a lot of anxiety and problem avoidance — and is likely to result in the tower of spaghetti collapsing when the marshmallow gets placed on top.

Sometimes the biggest challenge is not surmountable within the time constraints of a project, a semester, or even a year. In this case, the student has a choice: reach the highest level of progress they can and leave the innovation as a work-in-progress or take what they have already learned and apply it to a different project. Both options should be available to the student. As long as the project is not centered around the end product, both paths will allow the student to practice the skills of innovation.

Problem Solving

Problem solving in high school usually involves finding answers that already exist. Labs in science class are not true experiments; they are recipes for scientific processes that were solved long ago. Problems in math are not new applications of mathematical concepts; they are imitation projects for practicing math skills. In subjects like English, there are many standardized methods, like the five-paragraph essay, meant to take many problems out of the equation so students can focus on other skills. Art students often solve problems, but those problem-solving skills are not taught intentionally in the same way as techniques.

Problem solving is built into maker education. Often, we make things to solve a problem seen in the world. Projects take the form of design challenges requiring students to respond creatively to a problem presented to them. After every prototype, there are problems with the design that need to be fixed, worked around, or tinkered with. As with all of the other maker skills that go into innovation, students need to learn that there are problem-solving techniques they can learn in the makerspace, which can be applied to other areas of their lives.

There are many problem-solving frameworks that apply to different fields, but most of them boil down to three main steps:

1. The first step is to **clearly define the problem**. "It doesn't work, so we need to do something else" is a vague and unhelpful statement that doesn't advance the problem-solving process. It's essential to look at the problems that present themselves at the moment, not look down the road to possible issues for which we don't yet have information.
2. While working on the problem statement, students need to **challenge their assumptions**. There are many ways to do this, but the easiest is to get outside feedback. False-Consensus Bias will lead us to assume that others will see the same problems, challenges, and solutions as we do. Feedback from teachers, classmates, and others outside the class can help broaden the possible ideas that the students can draw upon.
3. The next step is to **decide on a strategy to solve the problem**. A solution can be arrived at in many ways, and the problem statement will determine them. Sometimes there will be several viable strategies, and it will be up to the student to decide which one they will

use. If students are working in a group, they might divide these options into parallel paths, but if they are working independently, it's better to pick one rather than divide their attention.

The most common solution strategy is **research**. There are a lot of avenues for research, from finding an expert to talk to, getting feedback from a larger number of people, and finding online videos and websites from reputable sources, to conducting your own tests. Many students will turn to YouTube and can end up going down the rabbit hole of videos without much helpful information to show for it.

Another solution strategy is **tinkering**. Tinkering is the process of trying lots of small changes and adjustments to an object or process until it works. Much like online research, tinkering can be an aimless or targeted thing. Tinkering is often defined negatively, implying endless tweaking and adjusting with no apparent knowledge or goal in sight, but there are ways to tinker strategically.

A third solution strategy is **testing**. If the solution is a clear decision between two options, with a single variable that students can toggle, A/B testing can help them make decisions. The challenge with this solution strategy is that students often want to introduce too many variables into a testing phase rather than test one variable at a time. Collecting data is another way to test a solution. By comparing the strength of different materials or making several variations and comparing how they function, you can strategically arrive at a solution.

There are many ways to solve problems, but using these three solution strategies in particular can help get students thinking strategically about problems they run into during an innovation project. As the task becomes more complex, it can be hard to keep all these options in mind and know the next best step in the process. At this point, students need to step back and look at the big picture.

Project Management

At the beginning of an innovation project, there are more unknowns than knowns. It takes a few rapid prototypes and a little research, testing, and tinkering to start understanding your project's most significant challenges and necessary steps. Once students have enough information and feedback, they can begin laying out the project strategically.

Imitation projects are excellent for teaching students to pay attention to details, while innovation projects help students learn to step back and see a project as a whole. This strategic thinking is one of the most valuable and applicable skills students can practice. At this point in an innovation project, students go from executing a game plan to creating one.

It's helpful to get to this project stage as soon as possible, but attempting to lay out a plan too early will likely be a waste of time. The best way to know if a student is ready to lay out a strategic plan is if they can describe their project in a narrative format from beginning to end. Can they, in a written or verbal paragraph or two, complete the statement, "In order to create _________, I will need to learn_________, test out _________, and get_________ to work." If their response to that prompt indicates that they have a firm grasp of the project, they are ready to start laying out their strategy.

There are many diagrams, charts, and organizers to help create a strategic plan. All these methods involve subdividing tasks in some way. The end product is divided into major tasks, which are divided into smaller tasks, which are divided into lists of steps, and on and on depending on the complexity of the project. In chapter 11, I share my project plan diagram, which I only have students fill out towards the end stages of an innovation project.

Adam Savage, in his book *Every Tool's a Hammer*, lays out his strategic planning method:

- He starts with The Brain Dump, where he lists everything he can think of that might be necessary to complete it.
- After getting into the project, he adds The Big Chunks, which break the big idea into manageable pieces.
- The big chunks get itemized and prioritized, allowing Medium Chunks to be added as part of each big chunk, and so on.
- As big chunks get broken into manageable pieces, it becomes clearer which tasks should be done first and which saved for later.

This branching visual finds its way into many of the available project management tools. The phrase "workflow" is used as if a project is a series of creeks and rivers. Little details trickle into flowing tasks, which combine to form major challenges, which are overcome to create a mighty project. Everything is connected in one strategic plan.

Building strategic planning into your innovation projects helps ready your students to manage a complex and challenging task. Depending on their time and available resources, they may not be able to complete the object they have designed, but with this plan you can see that they *could* accomplish it. Innovation projects can take weeks, or they could take years, and it is not helpful for the growth of our students to limit their thinking to project/semester/school-year limits. If the student truly desires to complete the project, they will find a way to do it, even if it is years later. If maker education aims to help students learn to solve big problems, a well-designed strategic plan is both a tool for students and an assessment strategy for teachers.

6

MAKER COURSE NUTS & BOLTS

The creative process is not a procedure. If you have come this far with me and now hope I can lay out a day-by-day plan for a maker class that incorporates all the skills mentioned in the previous chapters, I may be about to disappoint you: There is no single way to prepare and structure a course where the proposed outcome is new and unexpected. To try would negate the value of the type of educational experience that maker education offers.

The Lost Interview is a 1995 documentary interview with Steve Jobs that was released in 2012 and can be watched online. This wide-ranging, one-hour interview was recorded after Jobs was forced out of Apple and before he returned to save the company from bankruptcy. In one portion, he describes the difference between what he calls process and content. According to Jobs:

- **"Process"** is the attempt by companies to institutionalize how great ideas come about.
- **"Content"** is the creative output of its employees.

He describes how IBM, one of the biggest computer companies in the world, looked at some of the great successes of their early days, formalized the process of how those successes came about, and then tried to recreate their success with those same procedures. Inevitably they failed to yield similar results, which contributed to the company's downfall.

Instead of institutionalizing the process, companies should focus on the creative content of productive employees and give them the support they need to do their work.

In education, the curriculum often takes the role of procedure (what Jobs calls process), and becomes the goal rather than the student outcomes (what Jobs would call content). The teacher becomes a recorder of numbers and reader of instructions, not an actual supporter of learning and growth. The students who succeed in these classrooms often don't ask questions and are content to complete tasks without necessarily understanding the point of them.

Though I can't give you a day-to-day set of instructions, like GPS navigation, I can give you a map and some useful landmarks so that you can find your way around. It might be more challenging, but it will lead to more meaningful experiences for you and your students.

Structuring Time

In a typical class, including most art classes, it's common to look at the number of weeks in a school year or semester, and break them into somewhat even sections based on the content that needs to be covered, giving each topic an allocated slot. Inevitably whatever is at the end of the year gets squeezed as the schedule gets pushed and reorganized.

There are a lot of opinions about the flaws in this method and different ways to address them, but the biggest challenge for our purposes in the makerspace is making time to deepen learning through practice. If we focused on one skill a week until the end of the course, students would only get exposed to the skill in a surface way and have little time to put it into practice.

In sports, athletes expect to have practices and games. Practice is broken up into drills and scrimmages. Drills are when athletes focus on a particular skill, perhaps footwork in soccer or route running in football. Scrimmages are more freeform than drills, but often the action is stopped for coaches to discuss some skill or strategy that comes into play. There is still some coaching when it's game time, but players focus on executing what they have practiced between games. Afterwards they review their performance and build that into the next practice.

This structure allows athletes to grow and improve after being exposed quickly to crucial skills. If the coach doesn't introduce a necessary skill until the last week of the season, the team won't do well, and the athletes will not have time to develop a level of mastery.

How I structure my time with my students looks more like a season of basketball than a semester of biology. A semester-long *Introduction to Making and Design* class would look like this:

Half to two-thirds of the course is devoted to practice.

- Introduce technical and creative process skills in 1- to 10-day projects.
- Reflect and document progress on those skills.
- Teach basic tool use and foundational knowledge in short imitation activities.
- Emphasize skills from the three modes of making. Students may not complete their projects during this time.

Half to one-third of the course is set aside for execution.

- Students apply the skills they have been practicing.
- They continue unfinished projects or select new projects to work on.

- They identify which mode of making their project falls into and use previously learned skills to contribute to their success.
- They use creative process skills to learn additional tools and techniques, or improve their skills, to accomplish their projects.
- The teacher introduces periodic activities and reminders about creative process skills to small groups or the whole class.

This time structure aims to bring all students up to a particular level of understanding and experience and then allows them to apply those skills. Not every student will have the same experience. Some may follow tutorials and develop their skills in imitation. Some may find a modification project they want to focus on, making several customized items to give as gifts. Others may prefer to develop their innovation ideas to see how far they can push their new ideas.

For a year-long class, I break the curriculum into thirds. During each section, we refresh specific skills and add more, providing the opportunity for practice and execution in each.

YEAR-LONG MAKER COURSE CURRICULUM (36-week course):

I. *Introduction to Making and Design* (12 weeks)

A. Imitation (2 weeks)

1. Finding ideas
2. Setting goals
3. Paying attention to details
4. Executing: Follow a tutorial

B. Modification (2 weeks)

1. Personalizing
2. Working with constraints
3. Upcycling
4. Executing: Personalize or remix an existing object

C. Innovation (2–3 weeks)

1. Generating ideas
2. Taking calculated risks
3. Prototyping and iteration
4. Executing: Create an innovative concept and prototype

D. Execution (5–6 weeks)

1. Making things in the three modes of Making
2. Documenting experiences and skills used to make progress

II. *Real World Design Challenge: Making for Social Good* (12 weeks)

A. Imitation: Produce an existing assistive device (2–3 weeks)

1. Working systematically
2. Learning from mistakes
3. Paying attention to details
4. Executing: Manufacture high-quality designs as a team

B. Modification: Personalize an existing assistive device (2–3 weeks)

1. Working within constraints
2. Using human-centered design
3. Remixing
4. Executing: Make an assistive device for a specific user

C. Innovation: Design a new assistive device (6–8 weeks)

1. Generating ideas
2. Prototyping and iteration
3. Solving problems
4. Managing the project
5. Executing: Work as a team to design an assistive device for a real client

III. *Advanced Making: Using Imitation and Modification to Support Innovation* (12 weeks)

A. Generate an idea (1 week)

1. Identifying personal interests and needs
2. Exploring the needs of others
3. Getting feedback to expand ideas into innovation

B. Produce initial rapid prototype (1 week)

C. Solve the biggest challenge (2 weeks)

D. Execute the innovation project (8 weeks)

1. Reflecting periodically
2. Evaluating progress
3. Exploring multiple iterations

This course outline aims to balance periods of directed activities to introduce skills, activities to quickly put them into practice, prompts to reflect and evaluate them, and lots of time to work on longer-term complex projects. In the final eight weeks of the course, students have ample time to draw on all the skills necessary to pursue an innovative and personally motivating project to a high level of completion.

Phase 1: Introduction to Making and Design

This phase of the course focuses on the three modes of making and building basic skills in both technical areas and the creative process. Each makerspace will have different options available, and students will have varying interests from year to year. I've had years when every student wanted to learn how to use power tools, and others when nobody did. This introductory section of the course is a great time to give quick lessons on the basic use of various devices.

The assignments given in this phase of the course, or the course as a whole if it is offered as a standalone, are based on the skills practiced, not the end products. (I go deeper into grading and assessment in chapter 8.) At this point, the most important thing is for students to gain some practical techniques in the process of making and designing objects so they can apply these skills to future projects. The projects themselves are a means of delivering the experience of practicing the skills.

Anyone new to thinking about different modes of making may need help understanding why it matters what mode their project falls into. I've written about this in previous chapters and hopefully laid out a compelling argument as to why this is a helpful framework, but for the sake of this course outline, here is the boiled-down version:

1. The different modes of making rely on specific skills and techniques to be successful. Strategies that make sense for one type of project may waste time and effort in another. For instance, if you try to remix an imitation project, you may not learn the skills that were the point of imitating a process in the first place. If you focus on details too early in the innovation process, you may never progress in developing something new and innovative.

2. Sometimes in the innovation process, we need to switch gears into imitation or modification to gain knowledge or understand an existing

component. Mentally reframing the challenge makes these short-term goals more efficient. Tinkering your way through a tutorial to learn something specific can take a long time and add a lot of distracting information.

3. Falling into the trap of thinking that only "innovation" is a worthwhile endeavor in a makerspace will lead to many lost opportunities for student growth and engagement. Learning from the amazing things others have created or how we can alter the world around us is just as crucial as creating innovative projects.

IMITATION LESSONS AND CONCEPTS

Periodically when you visit an art museum, you will see an artist set up with their easel, paints, and canvas in the middle of a crowded gallery, copying from an original work of art. This is not some poorly disguised attempt to create a forgery from the original; it is an attempt to learn technique from a master artist whose only record of production, in all likelihood, is embedded directly into their art.

Here are some of the critical experiences and lessons that students can learn through the creation of an imitation project.

Finding ideas and setting goals

I've already discussed some of the challenges and benefits for students in finding ideas and setting good goals. The first few days of a maker course should be devoted to these two skills. Selecting a good project for the right reason is at the heart of working smarter, not harder.

There are many ways to emphasize these skills with students, and I share the thinking routines that I use in chapter 11, but the goal is for students to be able to visualize the entire project before completely diving in. Many people will claim that jumping into the water is the best way, and this may be true in controlled situations where the range of possible results is fairly certain. A backyard pool in June may be chilly, but it's not going to be frigid or dangerously deep. This is equivalent to a project selected by a teacher that is within the likely range of a student's ability.

When the student selects a project, they could be jumping into a placid lake or a raging river. They should learn to make some observations, get the lay of the land, identify if they have help nearby, and dip their toes in a little.

Once they have a better sense of the project, they can start to set more meaningful goals — goals that are challenging, achievable, and measurable. It's important to have a chance to talk about the ideas and goals the student has selected before they get very far into the project. These activities are only valuable if the students have thoughtfully engaged in them. As fun as it is to jump right in, the project is more likely to be successful with a slower pace at the beginning.

Paying attention to details

YouTube is a fantastic resource for makers. The number of tutorials, DIY projects, and unboxing videos keeps growing. Gone are the days when *This Old House* or *The Joy of Painting* had to be watched at the slow and steady pace the producers decided was appropriate. You can now skip forward, pause, and rewind streaming videos at will. Your students will almost always skip anything that seems unnecessary (in their opinion!), often missing critical details. Missing details can also happen with printed instructions as students zip past information, such as which bolt or screw to use, and grab one that looks correct.

A quick activity that helps expose this tendency is to ask students to write down instructions from a given YouTube video, something that is between 5 and 10 minutes long and includes an explanation of the steps. Something like folding an origami animal or a paper airplane seems simple enough but is actually quite complex.

I remember being given the task of writing down step-by-step instructions on how to make a peanut butter and jelly sandwich in a writing class. When the teacher did exactly what we wrote, it never resulted in a successful sandwich. In a maker class, this type of activity suddenly has new relevance.

Watch the video and create the instruction set yourself and identify the critical steps and information in the video. You can make an EdPuzzle and insert open-ended responses at these points, or you can be ready to pause the video as you play it. EdPuzzle can be assigned without allowing students to skip, which is ideal for this activity's final part.

- First, give students the link to the video to watch and have them create their own set of instructions based on the video. Allow them to generate these instructions however they want, giving them slightly more time than the length of the video. Most students will start to skip around to be more "efficient."

- Next, compare the various lists that your students come up with. Some will be short and a few will be very long. Some may be out of order depending on how they moved through the video, and others will be missing critical steps.

- Finally, watch the video as a class, pausing it or using your EdPuzzle version to have students input their notes at the critical points. How they word their instructions may differ, but they should learn that slowing down and following the instructions in order will save them a lot of time and frustration when they have missed a critical step due to skipping instructions.

Tools and parts lists

As a kid, and I know I'm not unique, I took pride in not looking at the manual or setup instructions when presented with a new item to install or put together. Shoving your way through a process and learning by making mistakes is indeed a way to learn significant amounts of information about an object, and most kids pick this process up through the ordinary course of daily life. In a class setting, it's important to give them a different experience of managing a project.

Looking over a set of instructions, and double-checking your parts list and the tools required, is a great habit to build, particularly in a non-packaged project from the internet. It is easy to jump right into a LEGO kit project because you can assume that all the necessary parts are in the box, but if you are building something from an online tutorial, it is helpful to know if you have the supplies and tools on hand. Often a complex project is halted because of a missing tool or material.

If an online tutorial includes a parts list at the beginning, have students copy this into their digital portfolio or print a copy. Have them physically find the parts and check them off as they go. If the project doesn't include a parts list, have them complete the tutorial and create their list. If they've learned to slow down as they go through the project, they should be able to identify all the needed supplies and tools.

Teaching students the value of looking over a project and evaluating it based on skills, tools, and materials will enable them to start envisioning the project before they even begin. This habit can allow students to figure out if they can get the missing supplies or find an alternate project that meets the needs of what they want to create but utilizes materials and skills they already have. This skill will come into play when students move on to more complex projects.

Thinking in steps

One thing that keeps students from challenging themselves is seeing an example of an end result and immediately deciding they don't have the ability to accomplish the task. The flip side is that students have unrealistic expectations of what they can achieve based on limited skills and materials. The ability to break things down into steps can help the former gain confidence when they see that each step is doable and forces the latter to confront the fact that they need to gain more skills before working on an unrealistic project.

Students often try to "brute force" their way through a series of steps, using trial and error to complete an assembly. There are always some things that are clear just from looking at the parts, and they can see that "this" fits in "that" and jump right to that step. Often, however, there is some intervening step that allows "that" to not fall back out of "this," and to skip that step usually means, in the best of circumstances, spending time undoing the process back to that point or, in the more typical cases, ending up with an inferior or incomplete product.

Storyboarding is an activity that can help students think about the steps involved in their projects. A storyboard is like a comic strip that describes a video or story sequence. By inserting images with captions, students can start to think through how all of the steps of the project fit together, and if there are any missing parts, they can slip those images into the order of the storyboard. Google Slides or PowerPoint are good all-purpose tools for creating storyboards, and there are also online storyboard creators. If students work from a YouTube video, they can grab screenshots of the critical steps to insert into the storyboard and add captions to explain them.

Encouraging students to think visually in their documentation can pay off in the long run. To reinforce this, I like to show videos from the YouTube channel Totally Handy. This channel shows unique projects in amazing videos that include no spoken directions. The thought that goes into every shot and how it demonstrates the step of the process is remarkable, and I have learned many tips and tricks by watching them.

Learning to break existing projects into logical steps helps students think through later projects of their own design more systematically. Such projects often involve a certain amount of trial and error as they work through possible solutions. If they can shut down unproductive paths and increase productive options, they can save time and get a better result.

Follow a tutorial

The sharing aspect of the maker community is incredibly valuable to maker educators, but it can lead to some pitfalls. Anyone can post their DIY project on YouTube or share a set of instructions on Instructables, and often these projects are difficult, or nearly impossible, to follow. To make YouTube videos that are less than five minutes, many steps get glossed over. Instructables has hundreds of thousands of projects but no indication of the difficulty level other than what the project creator may have added to their description. Only a small percentage who post projects are educators, and many fall victim to False-Consensus Bias.

The best instruction sets for new makers hold no assumptions about the maker's skills, instead laying out seemingly obvious information in quick reminders. They are highly visual, using as little text as possible, including easy-to-interpret symbols, diagrams, and well-photographed images.

Even when you find a great set of instructions, it may just be difficult to follow because the creator doesn't use the same phrasing, terminology, or tools that you have. The only way to get better at recognizing which tutorials will suit our background and abilities best is to evaluate a variety of them.

A simple discussion or reflection about what seemed good about the tutorial, its flaws, and how the creator might redo those parts is an excellent activity for students. Over time they will learn to start reviewing a video or tutorial before spending days or weeks trying to follow it.

Projects with imperfect or incomplete steps usually cease to be imitation projects and often become something else: abandoned projects. Learning to evaluate the quality of instructions helps students select projects they can accomplish and learn from and helps them document their work and write their own instructions in the future, turning them from content consumers to content creators.

MODIFICATION LESSONS AND CONCEPTS

Once students have gained some basic skills and have gone through a tutorial, they should be ready to apply those skills to personalizing a modification project. I suggest that students use the same tools or skills from their imitation project so they know how to apply them. Examples of personalizing an imitation project might include:

- If a student made a vinyl decal on the vinyl cutter
    - They can make a vinyl skin that fits their phone or laptop
- If a student 3D-printed something they found on Printables
    - They can remix a model from Printables on Tinkercad and then print it
- If a student used the laser cutter to engrave and cut a nameplate
    - They can cut a box from MakerCase.com and then engrave the lid
- If a student cut their initial using a scroll saw and sander
    - They can cut the parts for a wooden shelf

If students have prior knowledge, or if you offer a variety of quick tutorials during the imitation phase of the course, they will have more options. Here are some activities and lessons you can use to discuss the various skills that come into play during modification projects:

Working with constraints

YouTube is full of IKEA hacking videos. In these videos, makers at all levels turn stock IKEA pieces into custom designs by altering, adding, and modifying the design. Finding the best redesign videos to show your students is unnecessary. Finding videos where the maker walks you through creating their hack is better. This is because you can discuss the decisions made and what could have been improved or altered.

While it would be fun to give each of your students a Kallax bookshelf kit and let them have at it, this is not possible from a cost, space, or facilitation standpoint. Instead, they can digitally remix a Kallax or Billy bookshelf in Tinkercad by finding scale models on Thingiverse and having students redesign them.

When you present this challenge to your students, they should have constraints put upon them that force them to think creatively. Students may turn their basic bookshelf into a spaceship or teleportation unit if they are not constrained in some way. The goal is for them to modify and personalize, not redefine the original object altogether.

Some constraints you might consider:

- It started as furniture, so it should stay as furniture, just better and more interesting.
- Additions to it should represent things students could purchase, or make from items purchased, at a hardware store.
- The cost of the final project (not including the original piece) should be less than $200. Students can post links to what they find online.
- They must interview a friend or parent and then hack the shelf to that person's preferences.

Another option to engage your students would be to 3D print mini Kallax units and have students modify them with the materials you have available in the makerspace. This version of the activity gives you the opportunity for a miniature IKEA hack showcase.

Cataloging constraints

Working with existing designs and functions involves many constraints, and many seemingly obvious items are never considered. As students select a modification project, have them catalog every constraint that they can identify. Sometimes listing something as a constraint opens the possibilities of how far students can push that constraint in exciting ways or new ways of dealing with it. Turning a limitation into a desired feature can feel particularly satisfying.

Most constraints fall into a few broad categories, which can help students think about their project:

- Physical attributes
- Cost
- Function
- Ease of use
- Aesthetics

Next, they should highlight the constraints that will determine how the project should proceed. For instance, if a student wants to design a custom vinyl decoration for their cell phone, the material is unlikely to be a significant consideration since vinyl adheres to metal and plastic equally well. The physical attributes of size and shape are much more critical as the design must fit properly. More thought and time will need to go into solving those problems than how well it will stick.

Take apart/fix-it projects

Modification is about working with real-world objects, and gaining some deeper understanding about or mastery over them. Knowing how something works means you can modify it in a way that doesn't unintentionally destroy or lessen the original object.

Taking something apart to better understand it is eye-opening to students. The world is full of electronic objects that consumers are not meant to fix. When a student looks inside an old phone or broken laptop, they get a glimpse into the level of design and engineering that takes place to make these items.

Fix-it or upcycling projects, even those as simple as replacing a laptop or phone battery, help students understand that these things are not impossible. There are tutorials online that will walk you through disassembling most brands of laptops so that you can install new components. If the IT department at your school has a broken device or two that they would otherwise just get rid of, save them the time and let your students take it apart. Do a little research before handing it over to ensure any safety hazards have been dealt with. For instance, some electronic devices have capacitors that can hold a charge for a long time.

If electronics are not quite your speed, buy simple toys like squirt guns or toy vehicles at the dollar store. Strategically break them and challenge your students to come up with repairs. Designing and 3D printing replacement parts is a great way to teach functional 3D design and fabrication. Morley Kert's video *Free 3D Printed Repairs for Strangers* on YouTube is an excellent example of this process in action. You can even put out a call to others at your school for those needing minor 3D-printed repairs to make it even more authentic.

Taking Pride and Ownership

The videos posted by IKEA hackers and 3D printers are often a business for content creators. That doesn't change the fact that most of them take pride in their work and want to show how they accomplished what they did. The IKEA effect makes us feel more attached to things we had a hand in creating or designing, so students should learn how to represent that pride and ownership in their documentation.

Taking clear and well-thought-out images is not just important for documenting a project; it's a skill that can be useful in all sorts of business applications, from Etsy shops to portfolio websites. Students often snap pictures or record videos with all sorts of things happening in the background, things that can be distracting or outright annoying. The fact that almost every student is walking around with a camera superior to every camera I ever used as a high school student makes it all the more frustrating.

There are many YouTube videos on smartphone product photography, but here are the essential tips to get your students to document their ideas at a high level:

- Find a space with good natural light. If the light is too strong, parchment paper or a thin white sheet over the light source can be used as a diffuser.
- Clear the background. If it's a wall, tape a large sheet of white paper on the wall and let it rest on a table in a smooth curve. Place the object to be photographed on top of the paper on a table or the floor.
- Use a piece of foam core or a slightly shiny white surface as a reflector on the opposite side of the object from the wall. This will lessen any harsh shadows.
- Use an inexpensive smartphone tripod to hold the phone steady during the shot. There are 3D printable phone stands and tripod attachments that can also do a good job.
- Get close, crop out most of the background, take detail images, and consider angles.

Have students post their high-quality photos or print and display them.

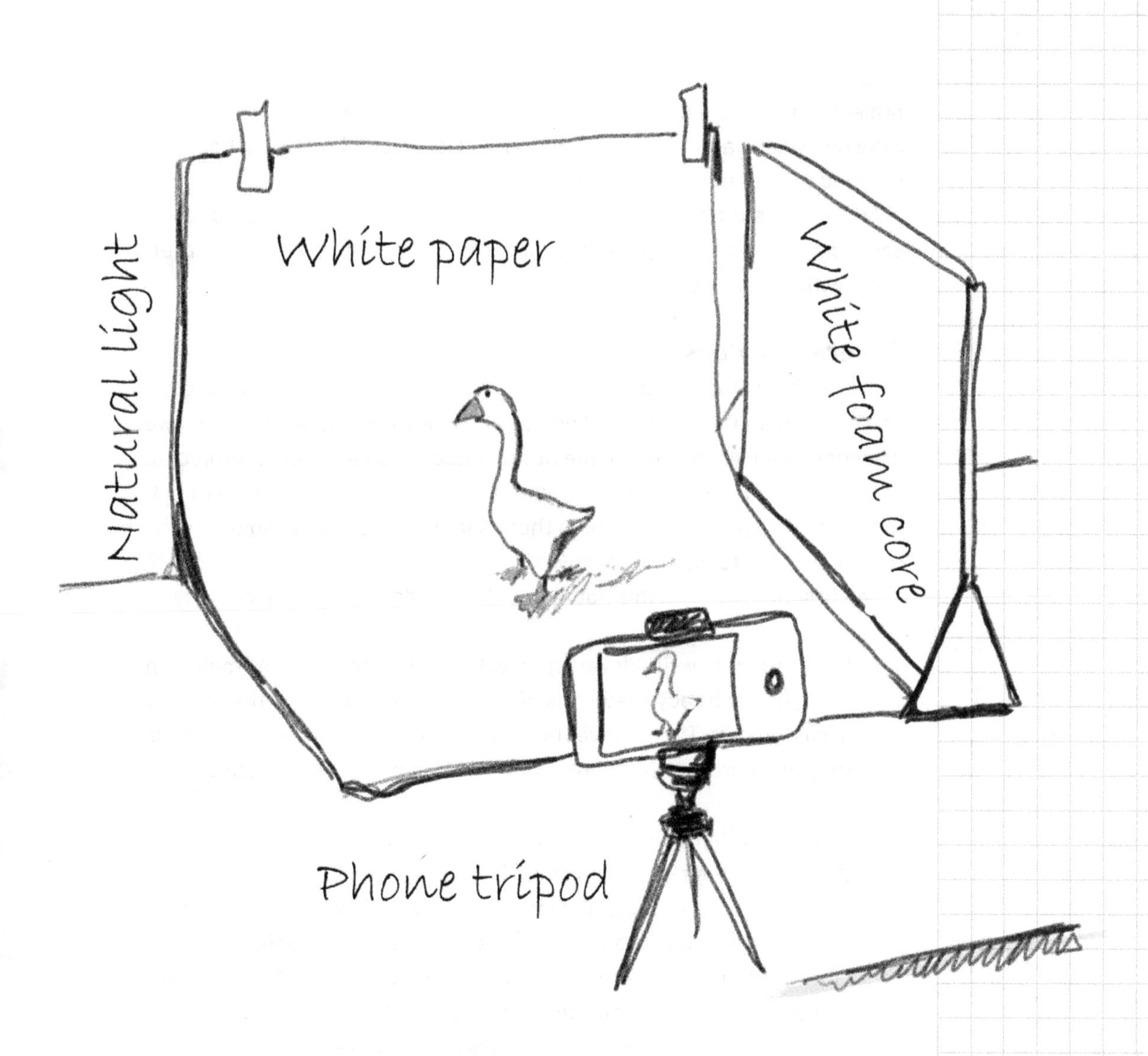

INNOVATION LESSONS AND CONCEPTS

In this first stab at innovation, you may only be able to peek behind the curtain with your students at the level of thought, effort, and engagement that they need to succeed with an innovation project. Again, the goal is to focus on the skills of the process rather than the result. While it is important that students at least reach the initial prototype phase, most will not fully realize their ideas beyond the conceptual stage.

Not completing a final product doesn't mean that they won't be accomplishing anything. A viable design with an initial prototype is the basis of every good innovation project and could lead to a completed project later in the course — or later in life. I have had students who returned years later and said, "Remember that idea I had? I've been working on it in my college courses."

To prepare students for an innovation project and all of the problem solving, strategic thinking, and resource gathering they will need, I prep them with a couple of activities.

Generating Ideas

I've already written about the value of "Yes, and" as a strategy to add to an idea to push it into new and potentially innovative territory. I have two activities that use this technique at the outset of an innovation project to help move students outside their comfort zone. These activities aim not to define their project but to help them see the possibilities so they can choose how far to push themselves.

The first activity is an individual one that we do on paper or digitally:

1. Have students write down an object that they could easily make. They could have already made this object, or know they have the skills to accomplish it. Their concept could be directly related to something they made in the imitation or modification phase of the course.

2. Next, have them add an element to this simple project that pushes it to the edge of their comfort zone. The feeling here should be, "I'm pretty sure I could make that." If a student's simple project was a 3D-printed character they found on Thingiverse, a project that pushes to the edge might be a customized version of the character they remix in Tinkercad. If they already did that as a modification project, they may feel it is possible to design their character from scratch.

3. Finally, have them add an element that pushes beyond their comfort zone. What can they add to the idea that is new to them? They may want to add LED lights to the figure to make its head light up. They may want to add articulated joints, so it is possible to pose the figure. Whatever they add should be something they have never done before but that works with the existing idea.

If the final idea seems interesting, encourage them to go for it. These sorts of projects will vary based on the experience level of the students, so it is helpful to talk a little with them about what they have done in the past. Both

"Yes, and ..."

Use the improv technique of "Yes, and" to generate a new idea and push yourself out of your comfort zone

1. Something I could make easily is ...	**2.** What can I add to this idea to push it to the edge of my comfort zone?	**3.** What is something I can add to the idea that pushes past my comfort zone?

3D models with embedded LEDs and articulated posable figures already exist and are not in and of themselves new. The skills a student will practice in creating their version of these things are the skills that will be needed when they set out to make something that is genuinely innovative.

The second activity I do with "Yes, and" involves the rest of the class and takes advantage of one of the best elements of the method: getting input from others. It can be easy to lock ourselves into our own perspective of how a project should look, and we don't consider other options. In this activity, we crowdsource ideas from others who have some of the same experiences as our own.

For this activity, I have students create a slide with several images representing their interests. These don't have to be projects they want to make, though that can be an option. Students might include sports teams they are passionate about, activities they enjoy, people that matter to them, places they like to go, etc. I then ask them to present their images to the class and talk a little about what the images represent about them. The rest of the class responds to a form with three questions:

- What is the name of the student presenting?
- Which of their four topics do you have a project idea for?
- What is an interesting project for them related to that topic?

In a class of 20 students, each student will get 19 ideas for projects that may be wildly different from what they might have come up with.

While they might not choose any of the suggested projects, and some might be pretty out there, they can consider if there are elements from any suggestions they want to incorporate into their project. It's good for students to learn that ideas don't have to come only from their own imaginations to have value.

Both idea-generating activities are designed to help students broaden their perspectives beyond their initial ideas and experience surprising ways to push themselves to consider new options.

Taking Stock and Stocking Up

Once students have decided on a project that is a stretch for them, incorporates some experiences they have already had, and involves something that feels new to them, it's time to start thinking strategically. Demonstrating the ability to think about the big picture is a critical element of their creative process.

I have used the metaphor of cooking in this book. Innovation in the kitchen occurs when we have enough basic skills and experience with ingredients, techniques, and tools that we can start to create new dishes. In keeping with this idea, I have students take stock of their current skills, materials, and knowledge that will assist in the project they've chosen and list what they need to stock up on to complete it.

Considering the light-up 3D figure from the previous lesson, the student might list the following items in their stock:

- Basic 3D printer use (knowledge/skill)
- 3D model of the figure (material)

To complete the project, they will need to add to their stock:

- Using Tinkercad to modify the model (skill)
- Learn how an LED works (knowledge)
- Connect an LED to a battery (skill)
- LED and battery pack (materials)

This document takes the form of a grocery list and will continue to evolve. Most students will probably be able to come up with the first two items to add to their stock but may need to learn what an LED is. If students know everything to add to their inventory, they probably aren't pushing themselves to try something new. To extend the metaphor, they will likely need to make multiple trips to the store.

Prototyping and iteration

Except in the highest profile circumstances like SpaceX's exploding starship tests, we never see the early versions of consumer products and electronics. Projects are kept tightly under wraps and the occasional "leak" of a new device becomes somewhat newsworthy. This can lead many students to feel like great innovations spring up fully formed out of the minds of creative geniuses.

The truth is that complex innovations often have humble beginnings. Sometimes it is based on cheap materials that get the look and feel of the product without the function. Sometimes it is based on existing products that get taken apart and recreated with new functions or capabilities. Sometimes it is a scale model, or a digital model. All of these options can act as a starting point to get from point A to point Z.

An interesting twist to the three modes of making is that imitation, which seems quick and easy, is all about slowing down. Innovation, which seems like it should be measured and slow, starts with just slapping something together. This is the time to pull out some cardboard, clay, foam, popsicle sticks, or whatever simple materials you have on hand to just quickly make something. A simple 3D model that can be printed the next day is another way to jump into the process, or mocking up a functional prototype out of existing components without regard to the physical constraints of the final design. With innovation, as long as it is based on experience with imitation and modification, it is a benefit to jump in.

The reason for this is simple. With imitation it is possible to see the whole project from the beginning. The steps are laid out and the processes are clear. Students should take the time to review all of those steps and processes ahead of time to see if there are any surprises or major obstacles. With innovation, it's all about surprises and major obstacles. The process of moving forward is how students will gain the knowledge they need to keep progressing. Without taking that first step, they can't gain the information and feedback they need for the next step, so they might as well get on with it.

Using what they learn with each prototype and improving the next is the process of iteration. The innovation project does not progress in a straight line. Sometimes an iteration may only be about one aspect of the overall idea, attempting to solve one of the problems without touching the rest. Learning to manage this process and create a strategic plan to solve these problems should be saved for the more advanced portion of the course.

What students need to learn during this phase is that prototyping and iteration are not the *result* of figuring things out, they are the *method* of figuring things out.

Pursue a Project

It is unrealistic to believe that students will design and execute a challenging innovation project in two to three weeks (innovation phase), so why not use all the remaining weeks to complete it? The answer to this is simple: They may have other things they want to do. Forcing them to focus on innovation is a biased view of what is most important in the makerspace. By this point, students have practiced many skills they can apply in the creative process. They have gained some good basic skills and knowledge and have done some hands-on work in all three modes of making. Let them use the space as they see fit in a productive project, regardless of the mode they want to work in.

They may choose to continue their innovation project, seeing it through to the end and even starting another one if they have the time. They may want to return to their modification project and make personalized items for their friends and family. They may want to learn more basic skills and complete one or several projects on Instructables. All of these should be options for this valuable phase of the course.

Up to this point, the teacher has directed a fair amount of the action, even though a teacher from a more traditional class setting might not see it that way. The students have been able to choose from a limited range of project options, and they have been guided significantly by the reflections and visible thinking routines. If the ultimate goal of maker education is to give students a feeling of agency and independence, they must have significant chunks of time (not days here and there) to apply what they have just practiced.

During this phase, the makerspace feels like a workshop, and the energy tends to be fun and joyful as students progress and experience minor or major successes. In chapter 7, I discuss several methods to manage this type of environment. In chapter 8, I discuss assessing and documenting student performance during this stage.

The subsequent two phases of the course — Real World Design Challenge and Advanced Making — follow a similar basic structure as this Introduction phase, with the teacher being a more active guide at the beginning and introducing new activities to students as new concepts become necessary. Sometimes I use the same activities described above to reinforce specific skills, but I also have other activities that are directly related to the main goals of these sections. Some of them are listed below and others are described in chapter 11.

Phase 2: Working with Clients

Students in high school and beyond should know they can meaningfully contribute to the world through making and digital fabrication. Many schools focus on entrepreneurship for designing products and starting businesses. While this is certainly a meaningful way to apply the skills learned in a maker program, I want students to think about how their ideas impact the world through the lens of service rather than business.

When athletes move up to a higher level of competition, they continue doing many of the same drills in practice. They do them faster, with more precision, and with more complex procedures. Shifting from the introductory portion of the course to higher levels represents an increase in expectations, not a complete change in skills explored.

The concepts and activities I use in the client-focused portion of the course are the same for any client-based project. It could be a design challenge from an international organization or a teacher down the hall who needs something for their classroom. This phase provides an opportunity to expand on the skills in the three modes.

QUALITY CONTROL WITH BETTER CHECKLISTS

Chapter 2 describes how checklists can be used in a step-by-step imitation process, and this is especially true when you are creating a design for a client. Organizations like Makers Making Change, e-NABLE, and Tikkun Olam Makers work with individual makers to produce and donate assistive devices that can be fabricated with digital tools.

If I make something for myself and it's glitchy, I can decide if I can live with the glitches. But when you are making something for someone else, quality control matters. Students need to learn how to manage the details of a client-based project.

I have students make checklists to facilitate a team of students 3D printing prosthetic parts, pill dispensers, or other devices. These checklists should be only five to nine items and will be rough to start. After the group has printed a device and figured out all the critical steps, they rewrite the checklist, focusing on clarity, and do it all again. What started as a READ-

DO checklist, DOing each step after READing, eventually becomes a DO-CONFIRM checklist, where students CONFIRM they did the steps they were supposed to DO.

Before

- ☐ Find a functional design
- ☐ Find high-quality files
- ☐ Download
- ☐ Clearly rename
- ☐ Put them in a folder
- ☐ Designate batches for each print job

Repeat

- ☐ Make sure designs show up on printer-computer
- ☐ Open design
- ☐ Make sure the project is sliced
- ☐ Remove supports
- ☐ Add brim
- ☐ Make sure all files are scaled the same (scaled to 130 on X, Y, and Z)
- ☐ Start printing and note down how much time it will take
- ☐ Double-check size
- ☐ Check back periodically to make sure there are no noodles
- ☐ Check back when finished at the corresponding time
- ☐ Cut/snap out and put it in a gallon-sized zip lock bag

PARTS LIST

	Print 1	Print 2	Print 3
Fingers	☑	☐	☐
TPins	☑	☐	☐
Pins	☑	☐	☐
Gauntlet	☑	☐	☐
Box	☑	☑	☑
Jig	☑	☑	☐
Phalanx	☑	☐	☐
Palm	☐	☑	☐

DEVELOPING EMPATHY

Working with clients requires understanding their needs and designing a solution that works for them. Usually, the client doesn't know what is possible, and the designers are sure they know the answer immediately. Both make assumptions about the problem, and students will only find the best solution once they build empathy for the client.

U.C. Berkeley's YPAR (Youth Participatory Action Research) Hub has tons of resources for helping students understand social justice issues through active methods. Sitting back and reading articles online is fine for learning the facts of a topic, but not for developing empathy. Many of the strategies on this site also apply to designing objects that help people solve problems. Two that I focus on with students are conducting interviews and doing observations. Some of the key elements we practice in interviewing are:

- **How do you create an open-ended question?** Open-ended questions cannot be answered with a "yes," "no," or a single-word answer. "Do you have breakfast in the morning?" and "What do you have for breakfast in the morning?" are not open-ended. "Tell me about your routine for breakfast" will yield much more information as the interviewee will mention the things that matter to them, which may only tangentially be related to what they eat.

- **Start with scripted questions, but follow-up questions are where you will learn the most valuable information.** The questions we write down before the interview represent what we know or think we know about the interviewee. If we want new or unexpected information, we must follow up on what they tell us in response to the scripted question. Students often try to run down a list of questions like a checklist, but to get to know the other person, we need to be active listeners and follow up on interesting things that come up along the way. I tell students that if they write 10 questions ahead of time but only get through three of them because they had a lot of follow-up conversations, they have had a successful interview.

To practice these skills, I have students interview each other to design something meaningful for their partner. Sometimes I give them a specific topic, like their morning routine, and sometimes they have wide-ranging conversations. If the prompt is to design something complicated, I have them bring it to a concept phase using the skills they have practiced ("Yes, and ...," cataloging constraints, etc.).

Getting the recipient's feedback is vital as a final step in this process. Feedback could be from a video response or a peer review form. It's important to stress that the project's goal is to design something that helps meet a need or solves a problem for the other person, so being open to critical feedback is valuable. Giving time for students to be able to redesign after this feedback stage can help them feel less pressure to get it right the first time.

After students have practiced their interviews with each other and received constructive feedback on how well they developed empathy and understanding with a classmate, they should be ready to interview an actual client. A procedure that I use to make this go well involves:

1. Work together as a class to come up with scripted questions for the interview.

2. Assign a few students to ask these questions during the interview.

3. Assign a few students to ask follow-up questions. They should be exercising their active listening to ask about critical points that pop up.

4. Assign some students to be note-takers. They should divide notes into general categories to focus on certain types of information (needs, problems, routines, habits, etc.).

5. Assign a student to be the moderator. They introduce the group, ask if there are follow-up questions, and call on the next person for a scripted question. They should thank the interviewee for their time.

These procedures help students understand how important the interview is for a successful project: Not just one that seems successful to the students, but is successful for the client.

Observations are also a valuable way of building empathy. A field trip is not always possible, but if it is, a visit to a local food bank, nursing home, or other service organization to do an observation can be a meaningful experience.

Observations can be combined with interviews, but the observations should be fairly unobtrusive. The goal is to see what actually goes on so that you can gain valuable information in order to make a better solution.

YPAR Hub has lesson plans around this, but they suggest looking for two types of information:

- **Assets:** Things that benefit or have value to the community or client
- **Issues:** Unmet needs of the community or client

Many unmet needs will be beyond what you can address in a high school maker class, but even small improvements that you can design may make a big difference in a person's life, or in the lives of those who work with a community.

It's possible to do observations without leaving the classroom. If the topic you are researching is common enough that people have documented their experience on YouTube, you can conduct observations from these videos. Have students pay attention to more than just what the subjects say: Look at the space around them, how they interact with items, how they interact with others. These visual cues can help you identify assets and issues even when you can't talk to the person directly.

Phase 3: Advanced Making and Design

By now, your students have a fair amount of experience in various creative process skills and have used digital fabrication tools in several simple and complicated tasks. They have experience working with a team on a complex design. They have created things based on personal interest, motivation, and empathy they have developed with a client.

Students have had an opportunity to experience and practice the skills laid out earlier in the book. But as much as learning the skills is important from the teacher's point of view, applying them will matter most to the students. The final third of the maker course outline is devoted to students pursuing a single project for an extended time.

When starting this phase, I like to show one of Adam Savage's *One Day Build* videos on YouTube. Many videos show a cleaned-up version of a project, which can be great for illustrating strategic thinking, but doesn't show the actual creative process. Adam Savage's *One Day Build* videos show the whole process, from thinking things out and solving problems along the way, to sometimes even messing up or getting cuts, splinters, or bruises. These videos will give students a sense of what the next several weeks may look like. In chapter 11, I share an activity that is centered around one of these videos. As we discuss and reflect on the various technical and creative skills Savage uses to complete the project, students are reminded that they have also developed some of these skills.

The goal of this phase is innovation: creating something new. In the tech industry, the word *innovation* often means game-changing designs. The iPhone was an innovation. Steel-reinforced concrete was an innovation. The steam engine was an innovation. It can seem to an outsider that these things were bolts from the blue, with no precedent, but that is never the case. The iPhone evolved from the iPod, the Newton, and other touch devices. The Egyptians reinforced their bricks with straw, and the first steam turbine was invented in the 1st century.

Innovation in makerspace projects can come in many forms. Some of them include:

- Significantly changing the form of an existing object to make it perform better or differently than the original.
- Taking an existing object and applying it to a new context or task. This often requires some redesign to make it work for the new function.
- Combining several existing objects to make a new object. This could also fit into either of the earlier categories.
- Changing the means of production or the materials used in creating an object. Using materials and methods that allow something to be DIY or significantly reduce cost are valuable innovations.

These elements are how innovation differs from modification. In modification, the object may be personalized or improved, but it essentially keeps the same form, function, materials, and means of production as the original. Water bottles with custom vinyl designs remain the same object as before the modification. A water bottle made out of carbon fiber is significantly different from the standard plastic or steel water bottle, so it constitutes an innovation. Through attempting such a project, students will have to explore how such a thing could be made, how to maximize the benefits of the product, and understand and minimize the drawbacks. These are innovation challenges.

Many issues can arise in pursuing such a project, and even though students have gained experience that can help them deal with challenges, they will need your guidance as they pursue these projects.

GENERATE AN IDEA

In the introductory phase of the course, students spent some time generating an idea. They will repeat this experience in this phase, but with higher stakes. While it is possible for students to pivot during this third phase of the course, they will likely not have enough time to change their idea once the 12 weeks are underway completely. Some of the questions to address before starting the project are:

- Is it an innovation project? If not, what can we add to make it into one?
- Is the goal of the project achievable? Does it rely on skills, tools, and materials that you either have, can acquire readily, or learn quickly?
- What are the constraints? Will it cost too much? Is there space to store it?
- How will we objectively measure success? What do we want it to be able to do?

These are all questions students explored earlier, but they need to have a handle on them before embarking on a project of this length and complexity.

If students want to do a project that is too complex or beyond the scope of the available options, have them think about a portion of the project that can be accomplished and is still challenging. If they want to build a DIY submersible that can take motion-activated photos of underwater creatures, focus them on underwater motion sensing as a first step in the project. Another option is to work in scale models. A collapsible, modular climbing wall can be built as a scale model that demonstrates all the functionality.

The goal is to end up with a challenging, achievable, and measurable innovation project that is interesting to the student and provides a lot of opportunities to explore, problem-solve, iterate, and eventually bring something new into the world.

PRODUCE A RAPID PROTOTYPE

Once the idea is solid, it is time to make the initial rapid prototype. Using all the prototyping skills and experiences from earlier, students should use the lowest tech materials possible to get an initial look and feel for the project. This first product can be cardboard, foam, hand-drawn images, wireframes on the computer, video, etc. The goal is to give them something to turn the abstract idea into a concrete object and get the creative juices flowing.

SOLVE THE BIGGEST CHALLENGE

As we have seen with the Marshmallow Challenge, addressing the most critical constraint and solving the most significant challenge should be done at the very beginning of the project. If these cannot be dealt with, there is no need to spend weeks working on the lesser portions of the project.

The most significant challenge may actually be unclear at first. The initial prototype can help make this clear. Learning a new technical skill may be the biggest challenge. Reaching a high level of mastery with a particular tool could be the thing upon which the project hinges. Sometimes there are multiple major challenges, in which case the teacher may need to guide the student to which task they should do first.

Solving this challenge may take a week, or it could take most of the time the student has to work on the project, but either way, it should be what they focus on before moving on to other parts of the project.

EXECUTE THE PROJECT

From here on out, the goal of the course is for students to execute the project to the highest level of completion possible. Some projects may not get there in the end, which is fine if the student has a lot of examples of applying the skills of imitation, modification, and innovation along the way.

This phase of the course is an energetic and engaging time, and it can feel exciting and chaotic. There are ways to manage the action in the makerspace, and there are ways to assess how your students are doing along the way. These are the topics of the following two chapters.

You may have noticed that what I've written about this final, critical third of the course is a fraction of the length of the earlier sections. This portion of the class is game time, and the coach stands on the sidelines and watches the players execute what they have practiced. The coach still has plenty to do: paying attention to the action, calling time-outs to give feedback, and occasionally calling plays. For the most part, though, the players make their own decisions in the moment in pursuit of their goals. Your job is to watch, encourage, remind them of what they have learned, and enjoy the show.

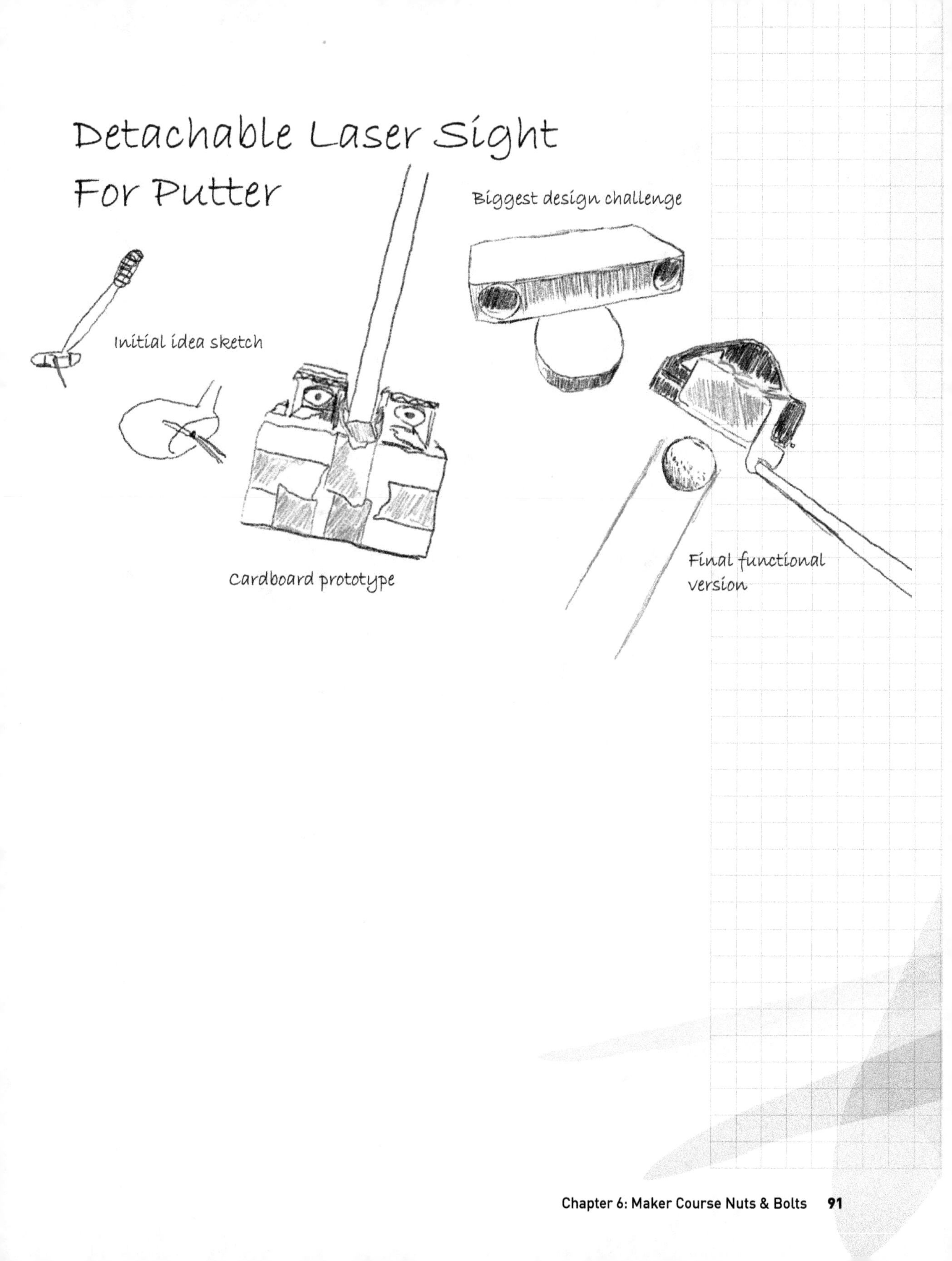
Detachable Laser Sight
For Putter
Biggest design challenge
Initial idea sketch
Cardboard prototype
Final functional
version

7

THE MODES OF MAKING IN PRACTICE

I've tried approaching this book like a project in the makerspace. First, I want to think about the big ideas of the project. What are the critical elements to consider? What should the goal of the project be? Once I've established some clear goals and feel like I have a handle on the big ideas, I start thinking about the skills I need to accomplish them. What techniques must I draw on to begin working strategically on the project? What methods should I use to organize myself and my tools? What are the most significant constraints of the project, and how should I make sure I can deal with them?

Eventually, there comes a time when the project kicks into gear. I would love to say that the prep and thinking create a situation where the project falls into place, but as Mike Tyson said, "Everybody has a plan until they get punched in the mouth." Inevitably something unexpected happens, and it can quickly feel like the project's challenges may be insurmountable. In an imitation project, the tutorial can be confusing, or missing information. Sometimes with a modification project, the materials don't perform the way you expect. In an innovation project, all sorts of unexpected issues can pop up.

From a teacher's perspective, this can be difficult to manage. Most classrooms are set up with the premise of teacher control. However, in a good maker class, the teacher gives the students a lot of control. The teacher sets the stage and tries to prepare the performers, but largely has to step back when the performance starts. It's up to the students to put into practice what they have learned. Later, the teacher steps back in to make sure the students evaluate and reflect on their performance to improve for the next time.

This chapter is about the nuts and bolts of running a productive, meaningful, and fun maker class. For years as a maker educator and an art teacher, I have worked with talented students and rookies, motivated and bored students, those with big ideas and the drive to make them happen, and students with no ideas who are afraid to take a first step. There is no single way to teach a course as regimented as math, and there is certainly no single way to run a class in the makerspace. Still, the following methods have helped me keep the creativity flowing — even after being punched in the mouth, metaphorically speaking.

Focused Chaos

A teacher that is used to students sitting at desks, looking at the board, and taking notes will look at the activity in an engaged maker class as a classroom management nightmare. A group of students might be cutting and gluing cardboard while others are watching YouTube videos, and the teacher may be engaged with one or two students on some power tool.

The traditional teacher may assume the video watchers are off task and the cardboard cutters are unsupervised. However, in a smoothly functioning maker class, the former are doing research, and the latter are executing a plan. The teacher is making sure the student with power tools is using them properly for safety and basic use.

On the other hand, it could also be kids off-task and wasting time while the teacher is distracted.

A good maker educator is the epitome of the "guide on the side" rather than the "sage on the stage." They set up the environment, both from a material and a culture standpoint, and then give students progressively greater amounts of freedom within that environment. The more functional the class is, the more things can happen within it. As students gain technical and creative skills, they will need less guidance from the teacher. There are many ways students can get information and help when the teacher is not working directly with them:

- Websites such as Instructables or YouTube
- Digital fabrication repositories like Thingiverse or Printables
- Information posted in the space about tools
- Other students with more experience
- Their own prior experiences from previous projects

Students who take the initiative to solve problems have gained agency over their work. They put in the effort not because of the grade they will receive at the end of the project, but because they have developed intrinsic motivation to get the job done.

I have had teachers tell me they would never let certain students loose around power tools, but those same students almost always do well when they have taken my classes. Most students who are disruptive are just bored. They don't stay bored for very long in an active makerspace.

It can take time for a student who has learned the habit of entertaining themselves through disruption to change those habits. They may have stopped seeing themselves as someone who can do well in school with the capacity to become an engaged student. In my 10 years working with students with learning differences and ADHD, I taught many kids who approached school as if they were going into battle. When they realize that building and designing objects does not rely on content memorization, test scores, fast reading, or similar skills that lead to success in traditional education, they start to see other types of learning at which they can excel.

But this depends on an environment where there is intentional teaching of skills. If you take a bunch of students and drop them into a makerspace and say, "Go for it," frustration, disruption, boredom, and the potential for injury become just as likely as creativity and innovation. Creating that intentional environment starts with a few basics.

Safety and Basic Use

This class routine involves teaching a tool or procedure's most basic functions and safety features. It often takes 1–2 class periods per tool and consists of having students quickly use the tool to produce a simple object while demonstrating safety procedures.

Maker education is not updated shop class. Shop class was vocational. It was intended to teach students technical skills and proficiency with tools, in preparation to become carpenters, mechanics, and metal fabricators. (For many reasons, shop classes have almost entirely disappeared in the U.S.) By comparison, maker education teaches students the design skills and creative-process skills to turn their own ideas into realities, not just execute and repair the ideas of others.

Maker education does not aim to teach students how 3D printers work. In a maker class, a student will almost certainly use a 3D printer, but the Cura slicer program has over 100 separate settings and adjustments that you can make. A typical student will likely need to know less than five of them to get the result that they want.

When I started working at my current school makerspace, I had a lot of experience generating ideas, helping students unlock their creativity, and working collaboratively with teachers. Still, there were a dozen complex tools in the space that I needed to gain experience with. I went to the Tech Shop, a for-profit makerspace chain, to take the safety and basic use (SBU) skills classes in CNC routing, welding, vacuum forming, and more. It was an eye-opening experience in how quickly you can learn what is needed to fulfill four essential requirements:

- The basic use of the tool
- How to avoid injuring yourself
- How to avoid injuring others
- How to avoid damaging the tool

In a few hours, I gained enough knowledge and hands-on experience to cut a simple shape on a Shopbot CNC router and to do it in a way that met the three essential safety requirements. Teachers are used to occasionally being only a little ahead of their students, and these courses did not make me an expert — far from it. They gave me the basic skills I needed to return to my own space and start building my skills with the tools I had. I have become proficient in some of them, like CNC routing. Others, like welding, which I only do occasionally with students, remain at a very basic level.

Safety and basic use projects are the essence of imitation. They are designed to very quickly give students a rudimentary understanding of how the tool functions and what it is capable of at the most basic level. These projects must be hands-on, so the information shared has a context in the real world. They should result in a finished object very quickly.

I only directly teach my students a few tools as a requirement. These are the tools most commonly used by almost all students in my classes:

- 3D printer
- Vinyl cutter
- X-Acto knife
- Hot glue gun
- Basic power tools (jigsaw, sander, drill)

Other tools that students can learn to use, if they're interested, include:

- Laser cutter/engraver
- Sewing machine
- Soldering iron
- Digital embroidery machine

Tools that usually are taught in a one-on-one situation are almost always project dependent, including:

- CNC router
- Floor tools (table saw, band saw, miter saw, etc.)
- Welder
- Waterjet cutter
- CNC milling machine

Sometimes these tools shift in priority as the interests and needs of my students shift.

In addition to these tools, I spend time teaching the fundamental use of:

- Simple 3D modeling (Tinkercad or something similar)
- 2D vector drawing (Corel Vector, Google Slides)
- Block coding for microcontrollers (micro:bit or Scratch)

Each One, Teach One

This phrase comes from the time of the U.S. slavery era. Slaves were denied education unless they needed it to perform their job. When a slave learned to read, it was seen as their duty to teach another — thus, each one would teach one. This sharing of knowledge is as important for students as it was for enslaved people to build community and look out for one another.

This class routine reinforces quickly learned procedures by having students teach each other what they just learned. Depending on the process and number of students involved, this routine could take one class or less. This routine combats the tendency of students to follow a series of steps without seriously thinking about or understanding them. When students need to assist another student with what they just learned, it helps reinforce the process from another perspective. Another benefit of this routine is freeing the teacher to work on other tasks while learning continues.

Everyone has experienced a feeling of being on autopilot when performing a set of rote instructions. It is possible to complete a complex task without ever really understanding or internalizing the purpose of the steps. Memorizing is not the same as learning; memorizing is a process, while learning is the acquisition of understanding.

When doing a safety and basic use project, or other imitation, students can easily perform a series of actions without understanding the point of them. This autopilot learning leads to needing to relearn the steps over and over because the students don't understand the point of the steps.

The only way to deal with this is to practice, and "each one, teach one" helps jump start that practice. The steps of this process are simple:

1. Students **LEARN** the steps of the process and the reasoning behind them: safety, basic use, or both.
2. Students **DO** the steps they have just learned with the guidance of someone who already knows and has done the procedure. For the first student, this would be the teacher.
3. Students **TEACH** the next student to go through the process steps, using only words and guidance, never touching the tool or pressing the buttons.

Teachers know that if you can teach someone a process, you remember the process better and are more likely to understand the purpose of the steps. Each One, Teach One very quickly puts students in a position to explain and oversee another learner who may need a firmer grasp of the concepts.

When students know this is coming, they are more likely to pay attention to better assist the next student.

- When there is only one tool, students will go through the process one at a time in a chain until the last student has completed the **DO** phase.
- If you have more than one tool, say multiple 3D printers, those students who have successfully helped others can help additional students.

During this process, the teacher should oversee the activity, ensuring that helpers are not doing things for their peers, monitoring that the helpers are giving correct guidance, and stepping in when unexpected circumstances arise. In the end, the teacher should act as the final DOer, pretending to be a novice that needs guidance, thus giving each student the experience of being a TEACHer.

This process not only reinforces the learning and understanding, but also distributes the knowledge and agency of all students in the class. Too often, a few students are picked and placed in the role of helper, which sets them apart from, and often above, the other students. This system can become a self-reinforcing dynamic that makes some students feel less engaged or appreciated. When all students feel that they can contribute to their own and others' success, they feel greater ownership and responsibility for the

learning. Not all students will take the initiative to fill this role, but every student should know they *can*.

Another benefit of this system is that while students are helping each other, the teacher can pay more attention to other areas. When students can go to fellow students as experts, they don't have to wait for valuable teacher time to start making progress. This environment is essential if the maker class hopes to offer individualized opportunities.

Individualized Imitation

Once students know a few options in the makerspace, they can start making more thoughtful choices about what they would like to make. In chapter 3, I spoke about the different skills students practice in the imitation mode, such as finding projects, setting goals, and working systematically. As a teacher, how can you structure these activities to allow students to gain experience in critical foundational skills?

First, don't give too much time on these projects. The goal is to learn how to follow a tutorial, not build something that will take weeks of work. That kind of endeavor can come later, and most novice makers aren't ready for anything more complex than 4–5 class periods.

Have students find a project or select one from a curated list. Before working, they should reflect on their goal setting and identify what will be challenging, what will help them achieve the goal, and how they will measure success. This reflection is how you, and they, will be alerted if their project is too difficult for them or so easy it is a waste of their time. A single class period spent considering these factors can save several on the back end. In chapter 11, you can see the visible thinking routine I use to get students to represent these factors in a clear visual way.

Every student may select a different project in this system; in fact, it's likely they will. Our students have individual interests, and the projects that seem valuable and interesting to them will be based on those qualities. The more we can let individual choice take place, the more engaged they will be in the process. That doesn't mean the teacher doesn't put a structure in place.

- **Structure 1: Time Limit** — These projects should be quick. Give students a week of classes, two at the most, to complete this first imitation project. What happens if they don't finish? Since the grade comes from the skills practiced, an unfinished product doesn't affect the assessment.

- **Structure 2: Clear Goals** — This process helps students think through the project's difficulty and whether it is achievable in the time provided. It only helps if students do this *before* they start working.

- **Structure 3: Identify Resources** — Students are used to thinking of the teacher as the source of information. Learned helplessness occurs when students feel unable to contribute to their own success. Help students understand that if the teacher is busy helping another student, there are lots of resources they can turn to:
  - Other students in the class
  - Reference materials, posters, and checklists in the makerspace
  - Websites, YouTube, or other online resources
  - Tinkering and testing on their own

- **Structure 4: Structure Teacher Time** — It is very common for the teacher to spend a lot of time with a few students and have a large number who, for various reasons, never get any attention or assistance. This situation can happen because students don't feel comfortable asking for help, don't think they need help, are less engaged in the project, or are making assumptions about their need for help. It's also possible for a teacher to tell someone you will help them in a minute and then get absorbed into another problem or task and forget to help that student.
  - Create a "Help List" on the board so students can put their names on it if they want to meet with you. Check the names off as you go down the list.
  - When you start the list, include the names of a few students you haven't met with in a while so everyone can meet with you and discuss how their project is going. Even if they don't need help, it's important to hear from them how things are going.
  - Keep track of students who did not get help and add them to the list for the beginning of the next day.
  - You can create two lists: one for quick help ("Where do I find the drill bits?") and one for longer help ("I don't know how to do something"). Knock out a few of the quick items between longer items, or direct them to helpers in the room.
  - If you notice certain tools trends for projects, say a lot of interest in power tools or laser cutting, offer a small group training on the

device. Use the Each One, Teach One method to work those students through the procedures.

- If you have students with more experience with specific tools, direct other students to them for assistance. Remind them to help, not do.
- **It can be tricky, but try to follow the Iron Rule: Don't do for others what they can do for themselves.** Remember that one of the goals is for students to gain skills, knowledge, and agency. One of the most satisfying things you can experience as a maker educator is hearing the words, "I figured it out on my own!"

Modification and Understanding the World

Modification projects allow students to learn about how their creations affect and are affected by the real world. By definition, they require that we interact with existing objects. To improve something, we first need to understand it. This mode of making is when measuring and diagrams come into the picture.

The activities and routines I use during this phase of the course are more technical than during the imitation phase. Students need to learn that numbers, measurements, plans, and diagrams take a little bit of time in order to save a lot of time. Some students will have to experience this the hard way, but if they can learn it after just one or two failed overnight prints, you will have taught them a valuable lesson.

HOW TO USE DIGITAL CALIPERS

Most students today do not know how to measure things. They may have seen a 1ft ruler, but a tape measure or a carpenter's square is about as new to them as a laser cutter. That's why one of the most valuable tools in a makerspace is a pair of digital calipers, and why teaching students how to use them is a worthwhile use of your time.

Digital calipers can make precise measurements in three different ways. The first is pinching an object between the large jaws to measure its outside width. The second is placing the small jaws inside an opening and spreading them out to measure the inner width. The final way uses a little rod extending from the calipers' end. You can push this rod inside an opening to determine the depth of a hole.

Most, if not all, design software is in the metric system. American students will have very little exposure to the metric system and will need a

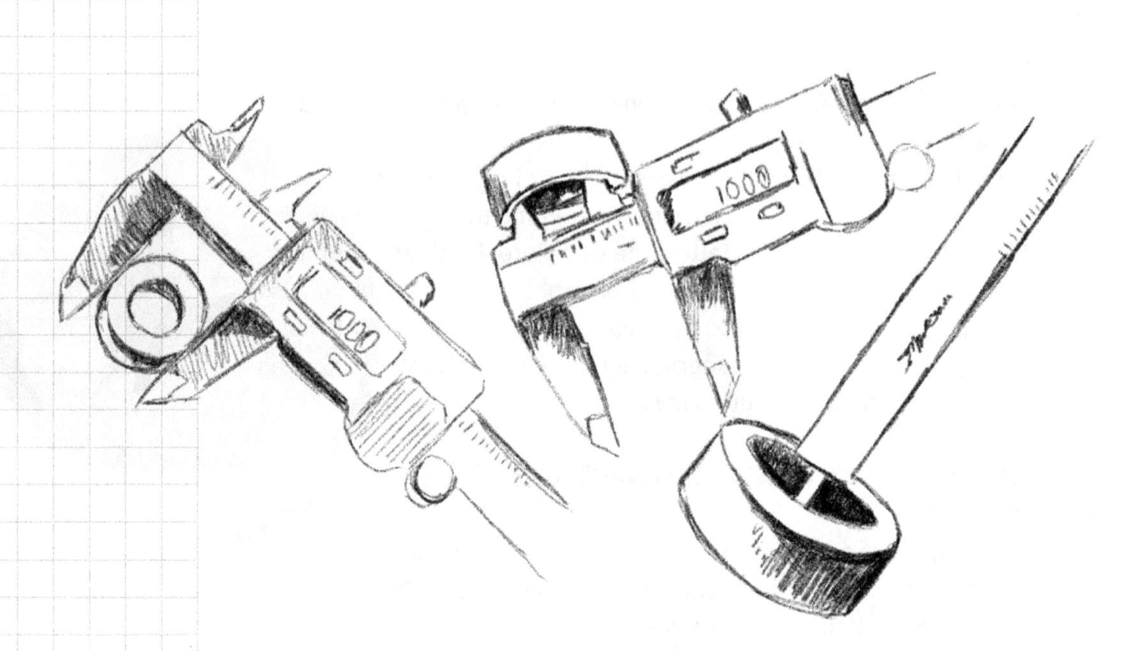

reminder about how it works. There are tape measures that have imperial and metric measurements, which can be very helpful.

Even though your students will likely have done some geometry in their math classes, they will probably have yet to realize that geometry applies to the real world. They may have an opportunity to use the Pythagorean theorem or a chance to use diagonal lines to find the center of a rectangle or square. Calculating the area of a piece of material is a simple process, but if all you've ever calculated before is a rectangle in a textbook, doing it in the real world can be an "aha!" moment for some students.

MODELS AND DIAGRAMS

To do most modification projects, students must fabricate the parts to execute their improvements. Fabrication might involve using tools as simple as scissors and X-Acto knives, but many will require creating a 2D or 3D digital file.

There are many advantages to digital fabrication:

- You can make multiple exact copies
- It's easier to be precise
- It can be quicker to produce the part
- It opens the world of digital design skills

When working with new or intermediate makers, I use the simplest design tool available to get the job done. I don't start by teaching students parametric 3D modeling programs like Fusion 360 or SolidWorks; I begin with Tinkercad, a free online CAD program that is more like building with blocks than making blueprints. For 2D vector drawing, I have students start with Google Slides or PowerPoint.

The reasoning comes from the course outline in the previous chapter and the law of diminishing returns. In the introductory section of the course, there are only two weeks to execute a modification project quickly, so students need to move from idea to design to the finished part as fast as possible. You can get the gist of Tinkercad or learn how to draw basic shapes in Google Slides in a single class period. This low floor can get students up and running extremely quickly on these early projects, which will serve as practice for skills they will develop later.

What are some quick projects students can do with these programs?

- Have students design a clip for their headphones that can hook onto a bag or a strap. Measure carefully and design it in Tinkercad, then print and test it.
- Create a vector drawing of a logo in Google Slides using straight lines, curves, and basic shapes. Vinyl cut, laser cut, or embroider their logo. Have them determine what their logo will go on so they have to measure the available space.
- Create a box or case to hold a specific object, like jewelry or electronics. Create the separate parts in PowerPoint with accurate measurements, taking the thickness of the material into account. Laser cut the parts.
- Create a 3D model of a specific object in Tinkercad, like a bracelet or pen holder, so that accessories can be designed for it. Use digital calipers to make measurements precise.
- Design a picture frame using Google Slides or PowerPoint. Make sure the opening fits the printed photograph, and then laser cut the design.
- Design a custom logo or recreate a team design in Google Slides, then cut it as a sticker to fit on a specific object, like a phone or laptop.

Each of these projects requires students to measure an existing object, create a relatively simple digital model that utilizes those measurements, and then see how those digital models can be turned into tangible things with digital fabrication tools, all without weeks or months of instruction in Fusion 360 or Adobe Illustrator.

MORE ADVANCED TOOLS

You should be prepared for most of your students to want to push the boundaries of what these tools can do. There is a point when Tinkercad and Google Slides can't create what students want anymore, and this is the right time to offer them more powerful programs. As before, students will need to utilize a lot of different resources to learn what they need to know to use them effectively, which may or may not include the teacher. If you have several students experiencing the same limitations, offer an introductory tutorial session to get them up and running.

I know enough about vector design and parametric 3D modeling to get students going, but I have many students who long ago surpassed my level of expertise. When students have a sense of agency and have the freedom to learn on their own how to accomplish projects that are exciting and meaningful to them, their capacity to gain new skills is incredible. No teacher could ever "stay a day ahead" of students in such an environment, nor should you try. There is always someone with more skills and knowledge than you; sometimes, it's a student in your class.

However, you should have a good working knowledge of some combination of these programs:

- **3D Parametric Design:** Design software that uses measuring and relationships between components to define objects. For instance, if I make a 10cm square with a button that is 1cm from the left edge, then I change the square to 8cm, the button stays 1cm from the left edge because that was the parameter I used to define it.
  - **Fusion 360:** Created by Autodesk, which also makes Tinkercad, Fusion 360 is free to students and educators.
  - **OnShape:** Cloud-based CAD program which offers free student accounts and accounts for hobbyists and non-commercial projects.
  - **Solidworks:** Industry standard software with special yearly pricing plans for students and educators.

- **2D Vector Design:** Vector design software creates shapes from points, called nodes, connected with lines, called paths. Digital fabrication tools like laser and vinyl cutters, CNC routers, and digital embroidery machines use these paths to cut and stitch.
  - **Inkscape:** Free, open source, vector design program that can be downloaded on most computers.
  - **CorelDraw/Adobe Illustrator:** These subscription-based programs are primarily designed for illustration, but can be used to create

complex vector designs for digital fabrication machines.

- **Corel Vector:** An online, subscription-based program with similar features to Inkscape. Utilizes cloud storage and the ability to share and collaborate on files.
- **Inventables Easel:** An online CNC design and manufacturing program which offers free educator accounts. You can design parts and cut them on a variety of CNC machines.

There are many more options with users who will advocate one over the other, and there will always be new programs. This list covers some of the best and most-established options with content to help you and your students get going with them. It's not necessary to be an expert in any of them, but as the teacher, you need to know a wide variety of resources to guide your students towards tools that will accomplish what they want them to do.

Other programs that might come in handy:

- **Vinyl cutter/Cricut software:** These machines usually have their own proprietary software that comes with the machine.
- **Digital embroidery software:** Though there are open-source options, like Ink Stitch, which is an add-on for Inkscape, purchased options will make the process much simpler.
- **CAM programs for laser cutters, CNC machines, and other digital fabrication tools:** Computer Aided Manufacturing (CAM) is a program that takes the CAD design files and turns it into a set of instructions for a specific machine. Programs like Fusion 360 can do both, but the learning curve for that is steep. Every new device has a new set of parameters. Most machines come with their own resources to learn how to work them properly. Teachers usually have to interpret these resources for students because they tend to be written for more experienced users.
- **3D printer slicers:** These are CAM programs for 3D printers. If your printers are open source, you have a variety of options to choose from. If not, your printer will have its own slicer program.

These programs rely on a healthy dose of numbers to do their work. Measuring, diagrams, and other specific features of materials are critical for them to produce successful objects. Students will have to learn — hopefully not the hard way — that paying attention to details and double-checking their numbers matter.

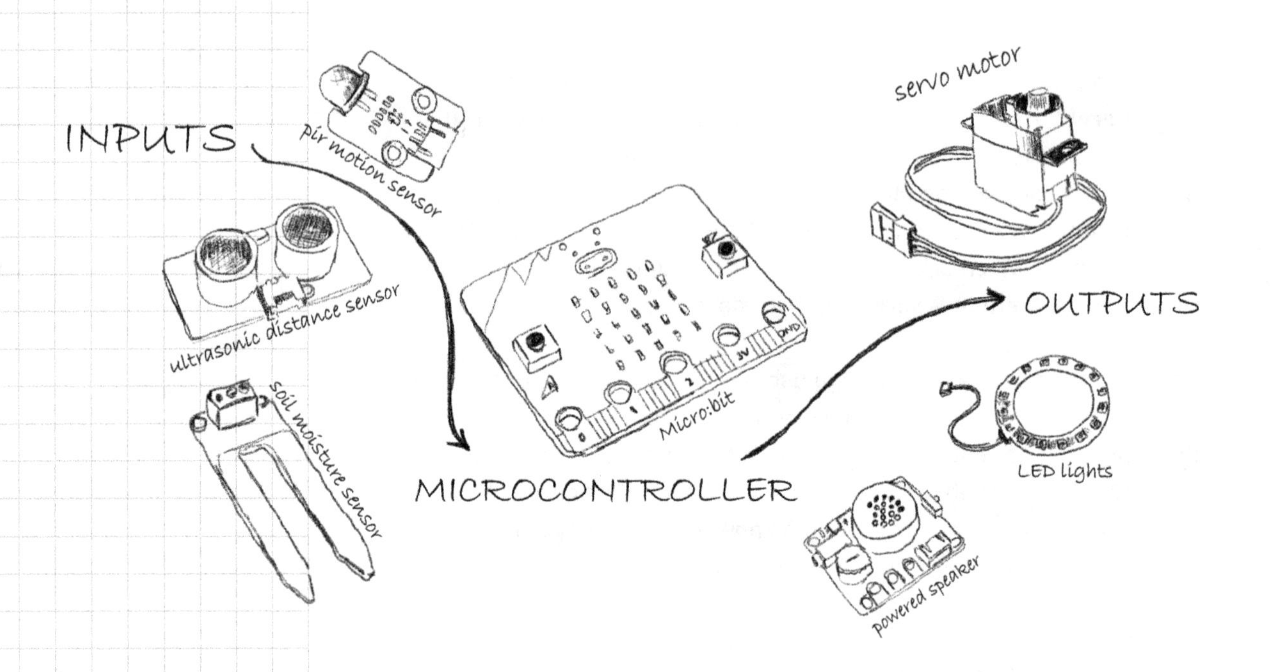

Coding microcontrollers is another area where it is important to have a good set of tools you can show students. Coding is easy to start but takes a long time to gain expertise. Most coding languages are better fits for your school's computer science classes, but being able to make things move, light up, make sounds, and perform other physical actions is both fun and valuable to makers.

Microcontrollers are miniature computers that students can use to execute the code they write, allowing them to interact with the physical world through digital instructions. There are dozens of different versions that are used in schools, and many excellent books that can teach you how they work and what you can do with them. The two most popular options are:

- **Arduino:** An open-source, highly customizable microcontroller. All sorts of sensors, motors, lights, and other attachments can be added.
- **micro:bit:** Small microcontrollers with built-in lights, speaker, sensors, and buttons, as well as the ability to attach other sensors, motors, and more.

In addition to wiring a microcontroller, students will need to be able to write code to tell it what to do. There are many different code languages, and they fall into two general categories: text-based code and block code. In text-based code, students need to know the syntax, the symbols, and how they need to be placed to make the code work. Block code systematizes text-based code into blocks which can be linked together to create complex sets of instructions.

Block code can do most of what students will want to do in the introductory phase of a maker course. Block code languages like **Scratch** or **Makecode** can execute complex instructions. By removing the likelihood of syntax errors, they allow students to focus on logical thinking and problem solving.

There may come a point when students want to do something that the blocks simply won't allow them to do. At this point, they will need to use text-based coding, which will offer more control. **Micropython** is a text-based language designed specifically for microcontrollers, and Arduino has its own version of the coding language **C++**.

It is rarely necessary to start from nothing with these sorts of projects. Examples of code to perform all sorts of processes with microcontrollers are posted all over the internet, and step-by-step tutorials abound. For the student who wants to use code for a project, not to become an expert in the code itself, it makes a lot of sense to modify existing code and to learn enough about it to get it to perform its function.

Innovation Through Collaboration

I've already discussed many strategies, such as rapid prototyping and problem-solving methods, to support students through an innovation project. The maker course outline provides many opportunities for students to gain basic skills and develop working strategies they can apply at various times when pursuing a new idea. Innovation challenges also present an opportunity for students to experience working as part of a team in a design challenge.

Design challenges come in varying levels of complexity and difficulty. They can range from simple prompts provided by the teacher meant to have students experience creative design, to challenges posed by international organizations with prizes and awards for the best submissions.

As a teacher, the biggest challenges of a team project are making sure that everyone contributes and being able to give assessments and feedback based on those contributions. We have all experienced group work in a school where one or two team members did the most work or the most challenging job, and the rest slacked off. Usually, there was a group grade on the project, and the students who worked the least benefited from the efforts of the hardest workers. In contrast, the motivated students were penalized by the laziness of the rest. I talk more about the assessment of group projects in chapter 8, but there are ways to make group projects more equitable and manageable by how they are structured.

Team members in the real world do not equally share responsibilities. They are divided up, and members have specialties that coordinate with one another. The goalie on a soccer team is not expected to take shots on goal. In a design firm, the person creating the 3D models will likely be someone other than the person making the ad campaign materials.

Whoever does what, team communication matters. No matter what their task is, students should be able to communicate and collaborate with the other members. At any time, the team members should know their project status, how their role plays into it, and what needs to come next.

There are many different roles and tasks that can be assigned that are of equal importance:

- **Concept Designer** — Makes sketches of possible solutions to the overall goal and challenges that come up through the overall design process.
- **CAD Designer** — Uses 2D or 3D modeling tools to create designs for prototypes that can be fabricated for the project.
- **Fabricator** — Uses the models to fabricate and assemble the parts.
- **Tester/Evaluator** — Evaluates the various prototypes and identifies issues to be addressed. Can get feedback from various users, but coordinates the process.
- **Documenter/Storyteller** — Documents all of the phases and versions and tells the story of the creation of the final design.
- **Project Manager** — Coordinates the various team members, identifies points where discussions need to happen, and keeps the ball rolling.

The significant tasks of these roles will occur at different times in the process, but in a good design challenge, there will be many cyclical iterations. A possible process may look like this:

- All group members throw out ideas in an initial brainstorm. The **Concept Designer** makes sketches of some of the most promising ideas.
- The **Fabricator** starts making basic cardboard prototypes based on these ideas. The **CAD Designer** starts to work on 3D models. The **Documenter** takes pictures of the various designs and starts to create a process video or project notebook.
- The **Tester** experiments with the cardboard prototypes and provides

feedback to the **CAD Designer**, who incorporates the input into the 3D models. The **Concept Designer** sketches out improvements to the design or creates new designs to address the issues of the prototype.
- The **Project Manager** communicates with all team members, passes information between members, and leads regular huddles to ensure everyone is on the same page. The **Documenter** takes notes, pictures, and videos of different elements to keep for later.
- This type of interaction continues until the project is completed or progress comes to a halt. If the project falters, the teacher needs to step in to identify the issue and restart progress.

It's pretty easy to figure out where the issues are when progress breaks down. If the Tester doesn't have anything to test, it could be that the Fabricator is inefficient in making new prototypes. If the Fabricator has no parts to produce, the CAD Designer may not be working very hard or need help. If the CAD Designer doesn't have any new designs to work on, the Concept Designer may not be passing on any new ideas, or the ideas may not be well sketched out. If nobody seems able to describe the current goal, the Project Manager is probably not handling communication well. If the portfolio or video telling the story is blank, the Documenter is either not recording or not building the story.

The teacher's role is to step in for periodic progress checks. The Documenter and Project Manager should be able to summarize and display the team's work, and the teacher can give feedback on what is working and what is yet to be accomplished. Some groups may need a lot of guidance from the teacher to keep moving, and other teams will only need a quick check-in. Waiting until the end of the project to find out how things are going is much too late and won't aid in the assessment of the project.

Anyone who has ever watched a basketball or football game may have noticed that the final few minutes of game time can take an extremely long amount of real time. At the end of a project, every decision matters, all of the fine details start to come into play, and time stretches out. The end result may be close, but you might have to make multiple versions with very slight differences, each of which takes as long to fabricate as the earlier versions. The last 10% of a project can sometimes take as long as the other 90%. Most students won't be prepared for this and will need some coaching to stick with it, but the satisfaction they get when that final version comes off the printer or cutter is the best motivation for future ventures.

The Self-Sustaining Makerspace

I said at the start of chapter 6 that I couldn't give you a step-by-step path to follow, but that I could provide landmarks and a map so you could chart your own course. Using these strategies — and adding other useful ones you come across as you become more experienced — will create an environment your students will roam around in. If you create an environment with only one path, your students will either walk it quickly or slowly or sit down and make no progress. This one-path learning environment is how most of their classes in traditional education work, and it will be a familiar, but not particularly valuable, experience.

If you create many paths, some clear and others unfinished, students will choose the options that appeal most to them. If you include guideposts and give students the tools they need to find their own paths, they will come up with new and exciting ways to use the space. The physical area and resources of the makerspace may be limited, but the combinations that students envision make the possibilities nearly endless.

It is essential to emphasize the skills that students demonstrate rather than the products they produce to encourage them to explore these possibilities enthusiastically. It's easy to look at the end product, check off a bunch of boxes on a checklist, and give a numerical grade. In order to support the open and creative environment that maker education offers, our evaluation and assessment methods need to reflect the goals of working in the three modes of making.

EVALUATION, DOCUMENTATION & REFLECTION

Now that we have a plan and some activities and lessons to implement, as well as some structures to support creativity and agency, the final piece needed to teach making in an intentional way is assessment. How do we know that our students are growing and improving in the skills laid out in this book? How do we know when they are pushing themselves to try new things, take risks, and learn when their attempts fail? How can we distinguish the flashy, easy project from the ugly but deeply challenging one?

There is no simple solution to these questions. No Scantron bubble sheet can tell us how our students are doing. In this, more than any other task, we need to learn the best lessons from the art teachers that have wrestled with these questions for generations.

There are many theories about how to assess this sort of education. We need to recognize that most teachers must fit their assessments into a point scale between 0 and 100%. This system has many obvious flaws, and if you don't have to use it, you should count yourself lucky, but for the rest of us, finding ways to give meaningful assessments that can still work with percentages is essential.

On the other hand, it is important to put only a little stock in that number. What your students will remember and take away from your class will be the projects they do, lessons that capture their imagination, and their experiences in pursuing their goals. The number may be what colleges and administrators want, and it will be nice if it aligns reasonably well with the student's actual performance.

In 20-plus years of teaching art and creativity, I have tried several strategies and used many tools to assess students. This chapter discusses what I have found to be the most meaningful and valuable methods, though I'm sure there are countless others I will want to try in the future.

Percentages

When I was in high school, teachers and students believed that the percentage was an objective performance measure. If I got an 85 in a class, I performed measurably worse than someone with an 89. I got fewer multiple-choice questions correct; I made more grammatical errors on my papers. Since everyone did the same assignments under the same conditions, it was (theoretically) an objective measure.

Of course, this was never true. If I was in a smaller class, I would likely do better. If I had ADHD, I would almost certainly do worse, while if I managed to do as well as my neurotypical classmate, there was no recognition that

I worked twice as hard. In my art class, the teacher gave us assignments that were meant to look very similar at the end, with little personal input. This system allowed him to hold the pieces up and rank them in order subjectively, with the best being 100 and everything else down from there.

If grades aim to predict future success in college and beyond, the hard-working ADHD student should be ranked higher than the student who easily remembers the formulas. Hard workers accustomed to struggling are generally more successful than those with good memory but no experience pushing their limits, given equal starting points.

The first thing I would say about grade percentages is that in the makerspace, the exact number doesn't matter, but the range does. Think about the basic expectations for a student in your class. They might include:

- They turn in all reflection and documentation assignments
- They complete x number of projects
- They can accomplish these things without needing the teacher at all phases to direct or push them
- They listen to and consider feedback

What grade would you give the student who met these expectations? Some might think the student should receive a 100%. If they don't turn in a reflection assignment, they will lose some points, or if a project is unfinished, they might drop a letter grade. Isn't doing everything on a checklist what 100% means?

I would give this student a "Meets Expectations" and decide where on the percentage scale that falls. This standard will be a school-by-school and program-by-program decision, but it should be clearly defined and built into the grading rubrics. Let's say you decide that Meets Expectations equals 80%; if all of the rows on a rubric are checked in that column, the resulting grade on the assignment should be 80%.

Thinking of basic expectations this way allows students to exceed expectations in a number of ways, based on their engagement, skill level, and willingness to challenge themselves. By not telling them exactly what exceeding expectations means, you avoid giving them a new checklist, which would undermine the goal of having them think for themselves. Students who exceed expectations on the above examples may demonstrate the following:

- They turn in reflection and documentation assignments with good details and thoughtful responses

- They completed x number of projects that were significantly challenging and forced them to try new things
- They took steps to solve problems using various resources besides the teacher
- They considered feedback thoughtfully and incorporated ideas that they felt had merit

Any student in my classes who performed in these ways deserves to earn Exceeds Expectations or an A in my book. Students who don't use their time wisely, are disruptive, or deliberately work below their ability level receive a Below Expectations ranking.

Thinking about performance in terms of these ranges can be used with traditional percentage grades, letter grades, or standards-based grading. It makes more sense with maker education activities, where students work on different projects, advance at different paces, and experience different challenges.

Rubrics

I've spent many hours of my life writing elegant rubrics that identified what creative skills a student exhibited, drawing distinctions between the quality of brushstrokes and the variety of colors. Inevitably there was a student whose work did not fit the description of excellence on the rubric and yet who clearly did excellent work. I could never fully capture in words what I was looking for.

I also created checklist-style rubrics. If a certain 10 elements were demonstrated in a project, it was an A. Inevitably a student would go down a rabbit hole and explore possibilities that weren't on my list, so asking them to go back and arbitrarily add elements to get the grade felt silly.

The nature of creative work is that it is unpredictable. Students will find ways to engage and excel in new and unexpected ways. Students will also find ways to disengage and waste time in new and creative ways. No rubric can fully lay out what an A, B, or D project will look like. No checklist of elements can include every valuable component that might be present in the finished project.

The best tool I have found thus far is the "single point rubric." The idea of the single point rubric is to identify the basic standard, which works well with the "Meets Expectations" range, and space is provided for feedback if the area doesn't meet the standard or exceeds the standard. It looks like the following table:

SINGLE POINT RUBRIC EXPLAINING EXPECTATIONS

	Exceeds Expectations	Meets Expectations	Areas to Improve	No Evidence
Challenging		You clearly describe 1–2 ways in which this project idea will be challenging.		
Achievable		You clearly explain 1–2 skills/resources/ materials you already possess that make this project achievable.		
Measurable		You describe 1–2 critical design features that can be objectively measured that represent success.		

When you assign points to the rubric, Meets Expectations will equal whatever number you have decided that represents, and the other columns can be assigned a number or a range based on the feedback. A submitted rubric might look like this next table:

COMPLETED SINGLE POINT RUBRIC WITH TEACHER FEEDBACK

	Exceeds Expectations 10 pts	Meets Expectations 8 pts	Areas to Improve 6 pts	No Evidence 0 pts
Challenging		You clearly describe 1–2 ways in which a project idea will be challenging.		
Achievable	*You have listed a lot of skills that you already have and referenced projects done that relate. It seems clear that the project will be achievable.*	You clearly explain 1–2 skills/resources/ materials you already possess that make this project achievable,		
Measurable		You describe 1–2 critical design features that can be objectively measured that represent success.	*The design features you have listed are not measurable. Durable is a good goal, but how do we measure it? How would you test the durability of this project?*	
	10 pts	8 pts	6 pts	24/30 pts

In this example the student met the expectations for a challenging goal, so there was no need to write any feedback. They exceeded the expectations for Achievable, and the feedback points out the reasons for this. They did not meet expectations for a measurable goal, and the feedback for how they could do better is included.

I like this system because it reflects the reality that there are many ways that students can meet or not meet the standard. Filling in the feedback gives students helpful information about how they can improve and in what ways they are excelling. It is also much easier to create since the teacher only needs to describe the Meets Expectations criteria. Nobody reads a 10-row, 5-column rubric with a paragraph in every box. Students skip all this and look at the number.

Of course, students will likely skip reading the feedback on these rubrics unless they can use the feedback you include to improve their performance. To make this work, we need to think about how deadlines work.

Flexible Grading

When I was in high school, the deadline was the deadline. You turned in the paper/project/test on the day that it was due, and then you moved on. My precalculus teacher had a retest for those who didn't reach a certain percentage that took place two days later. To a large extent, if you did poorly on an assessment, it was up to you to figure out what you did wrong. If you were in a math class where the content was built upon earlier skills, you gradually fell further behind as your foundation became shakier.

This grading system does not promote growth. It is meant to sort students according to their ability and to allow the teacher to keep up the pace to cover the content of the course. The final cumulative assessment shows whether students were able to keep up or not. The best students get into the best schools and get the best jobs. It was supposed to be a true meritocracy.

It has never worked that way, of course. Some students who were not natural performers had the time or resources to get help to keep up. Others who performed well in the class had issues with testing and bombed the final. Many students for whom this system worked well in high school failed in college in more project- or collaboration-based disciplines. Some leaders of successful companies dropped out of school altogether.

Most schools today offer some system to help a student catch up when the assessment shows they missed or didn't understand some content. This could be working with the teacher outside of class, a system of point recovery, or redoing work for partial or full credit. Math teachers have students show their steps to give credit for the portions the student understood and help them figure out the bits that are giving them trouble.

If what we are looking for in maker education is for students to develop skills that they can apply in the future to creative problems, these strategies are not enough. What happens if you teach a skill early in the semester and it doesn't click with the student until the end? What if their early projects have easy problems to solve, and they don't get a chance to put their problem-solving skills to work until later? Should they be penalized for not meeting the "due date" for problem solving?

Standards-based grading allows students to keep improving and working on a skill until they meet the standard. Some students will meet the standard right away, and others may get it at the end of the year. If both students can use the skill, is there any point distinguishing which one mastered it *first*?

Unfortunately, most high schools and post-secondary schools don't

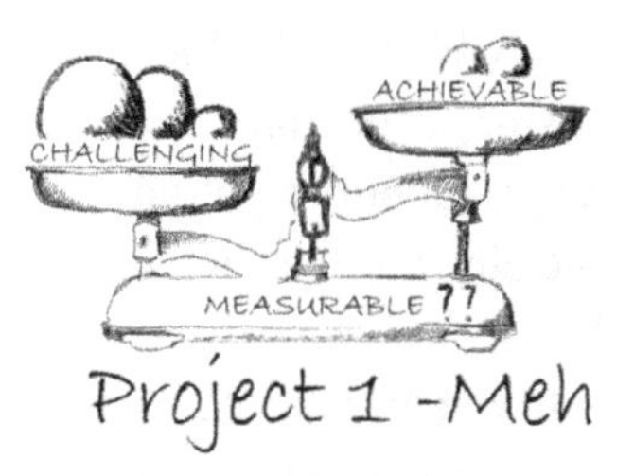

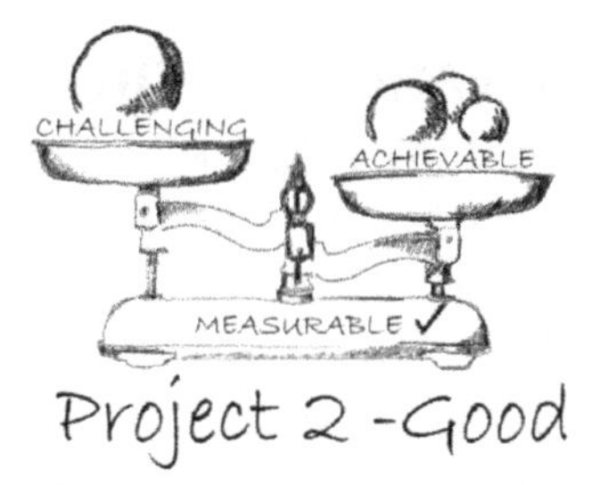

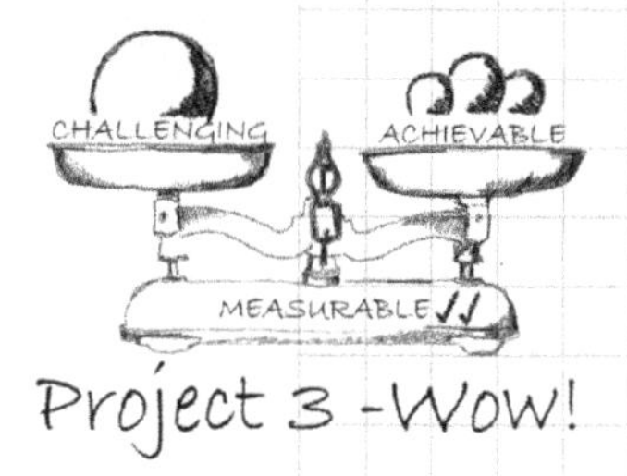

use standard-based grading, so it is necessary to think of the traditional grading system differently to reflect the reality of a maker class. Here are some grading practices that I have used:

DURING THE INTRODUCTION TO MAKING AND DESIGN PORTION

1. Skills are introduced in a logical sequence. Students are given an assignment to document or reflect upon their use. This documentation is accompanied by a single-point rubric that lists the basic expectations regarding the use of or reflection on the skill.
2. The rubric is filled out, and a grade is entered in the grade book. Students who excelled at using the skill received a high grade, and those who struggled to apply it or lacked engagement with it received lower grades accordingly.
3. As the introductory phase continues, students can resubmit the assignment when they have new work to document or new experiences to reflect on. Those students with lower grades will have ample opportunity to put these skills into practice on later projects. There is no penalty for resubmitting, as it reflects students continuing to work towards mastery of the skills.

DURING THE WORKING WITH CLIENTS PORTION

1. This portion of the class is mixed. Client-based projects need deadlines to meet the client's needs, but students are still developing skills as makers. Skill-based reflections follow the process described above, while product-based assignments with documentation have stricter deadlines.
2. Penalties for late work reflect the reality of needing to meet deliverable deadlines.

DURING THE ADVANCED MAKING AND DESIGN PORTION

1. With exposure to the skills and time to gain mastery over them, the assignments in this portion are more often process documentation. Hitting benchmarks to get projects going, especially in the early phases, is important so deadlines are applied.
2. Once projects reach the execution phase, students will give regular updates on what is happening and the skills they utilize to complete the project. These status updates should also meet deadlines, but with some flexibility built in to recognize that projects will move at different paces.

What Gets Graded

In any complex process, things can go wrong, even for the most prepared and experienced maker. One project cannot show a teacher how well a student can do all similar projects. If grading the finished product is not a good assessment, what is the maker educator looking for to assess student progress? I rely on three assessment tools for students in my class:

1. **Visible thinking routines** — These activities are tied to specific skills such as goal setting, creating a strategic plan, identifying constraints, and others that will benefit from some thought beforehand. These activities gauge a student's ability to predict some of the project's key elements before physical making starts. The visual nature of the activity makes it easier for the teacher to gauge the sophistication of the students' thinking as text, visuals, and symbols all contribute meaning and context.

2. **Reflections** — These activities ask students to think about what happened. What problem-solving strategies were used? What was successful? What still needs to be done? These reflections are done at the end of a project and at various points along the way so that students can apply insights gained to the continuation of the project.

3. **Process documentation** — Students should be able to tell the story of their project. What challenges have they been working on? What skills and strategies have they used? What have they needed to develop to move forward?

It may seem backward, but in a real way, I have my students make things so they will have something to reflect on. John Dewey said, "We do not learn from experience; we learn from reflecting on experience." If we only made objects, students would reflect on their experiences randomly and unintentionally. It would be impossible to say exactly what they learned from the experience. By reflecting on the process of creation, we, as teachers, can be more intentional about what students gain from their experiences.

Reflections

There are a lot of meaningless reflection assignments, especially in upper grades. By the time students reach high school, they have learned the buzzwords they must include in a reflection to get a decent grade. Reflections in many classes become ungraded exercises in doing what your teacher tells you, further reinforcing the transactional nature of the classroom with an activity that should encourage students to actually reflect on their experience.

A classic reflection prompt might look something like this:

Reflect back upon what you have done in this project. What are you most proud of? What were some challenges, and how did you deal with them? Did you give your best effort on this assignment?

These questions work fine in a one-on-one conversation, where the teacher can dig deeper into the responses. Too often, these questions are given as a written response assignment with little to no follow-up.

Students already good at deep, reflective thinking will respond with thoughtful answers to these questions. Students who need to improve at reflective thinking or expressing themselves in writing will probably write a sentence or two that is essentially meaningless. So, then, what should be a valuable exercise to get students thinking about their actions and give the teacher insight into how students apply creative skills?

For the last decade, Harvard's Project Zero has been sharing strategies that rely on visible thinking. The book *Making Thinking Visible*, as well as the Project Zero website, are excellent resources for the maker educator. Visible thinking routines and reflection prompts that require students to think and respond through visuals, symbols, and representations will make it easier for the teacher to get better insight into students' thinking. Are they considering options or taking the first idea that comes along? Are they invested in the project or doing something they think will be easy?

I have already described some of the visible thinking routines I use, such as storyboarding, to see how well students think about their projects. In chapter 11, I share some specifics that I have used repeatedly. Every year I tweak, improve, and update my routines to try and make them quicker, easier, and better at showing students' thoughts. If students can share critical information about their project in a 5-minute response, they have more time to implement that thinking.

Visible thinking routines work best for skills that rely on thinking ahead, such as setting goals, while reflection after action works best when there has been some concrete experience, such as problem solving. It is valuable to reflect on goals after the project is over as a post mortem, but that should never be the first time students think about their goals in a concrete way.

Visible thinking routine skills:

- Finding ideas
- Setting goals
- Generating ideas
- Personalizing
- Working with constraints

Reflection skills:

- Solving problems
- Learning from mistakes
- Working systematically
- Prototyping and iteration

Skills that work for both:

- Paying attention to details
- Remixing
- Upcycling
- Human-centered design

Having students reflect on their actions encourages them to learn valuable lessons from the choices they make and the results of those choices. Having them reflect in a way that asks them to do more than string a few words together that sound like reflection gives the teacher an authentic look into how deeply they are thinking about their process.

Feedback

The teacher must provide feedback when it becomes clear that students are not thinking deeply about their process and are therefore not learning beneficial lessons from their experiences.

Feedback takes time. In a maker course, there won't be much homework. There won't be papers to read or tests to grade, but there will be lots of feedback to give. You will write comments, record audio and video feedback, and meet with students during and after class hours. This is where the bulk of your time will go as a teacher.

Just as students should be able to express their thoughts visually to quickly and clearly represent them, teachers should work to give clear and concise feedback that students can benefit from quickly.

TYPES OF FEEDBACK

- **Face-to-face feedback** is best when students are missing some concepts. This situation requires a conversation so that you can determine the problem. Is there some underlying misunderstanding? Is the student motivated? Sitting down with the student, you can actively troubleshoot the situation.

- **Written comment feedback** works best for anything that can be expressed in a sentence or two, and they are usually about concrete actions. The makerspace is not a writing class, so the hustle and bustle of the space can make it hard for students to read and internalize longform feedback. Even students who are good at this in other areas will be in a more visual/active mode in the makerspace.

- **Video/audio feedback** is good for sharing ideas from the teacher to the student. You are giving thoughts and ideas that they may or may not take action on. Everyone tends to ramble on camera, so be concise and express yourself clearly.

- **Screen recording feedback** or recording yourself talking while showing what is on your computer screen, is one of the best methods for giving technical assistance remotely. Using this method, you can highlight photos, sketch diagrams, or bring up websites and videos while discussing what you want them to learn. This method is a great way to give concrete details and general ideas you want to share.

Unless you are pointing out some missing element that students need to complete, responding to feedback should not be attached to grades. Feedback is coaching, and you won't always know what students have learned from your coaching until they apply it to future situations. Sometimes it might feel like you are talking into a void, but when you see students put your feedback into action, you'll know they are listening.

Process Versus Product Portfolios

As an art student, I had multiple large folders for storing drawings and paintings. As an art teacher, I repeated this ritual for my students. I was convinced their creative process would be evident in the final product. Undoubtedly, some of it was visible, but much of the thinking students engaged in during their creative process was only visible if they asked questions or I made it a point to check in with them.

There is no perfect way to know everything a student goes through in a creative process. Even asking them what they think at a particular moment will likely result in a surface-level explanation. Humans don't process our actions like sports commentators ready to offer an analysis of every move. We do things and later process what we did into an explanation. These explanations help us internalize what we did and do it better next time.

I no longer grade the final product; I grade students' documentation of their process. As I have pointed out, a student can create a very technical and impressive-looking project without much deep thought. In contrast, another student's failed mess exemplifies taking risks and doggedly pursuing a big challenge.

At the same time, if a student has a weak project but beautiful documentation, their grade should reflect that. The rubrics for documentation are only partly about what is included in the portfolio and visible thinking routines. The rubrics include things like:

- There is evidence that you tried something new
- The weekly plan includes a challenging amount of work
- There is evidence you tried multiple methods to solve the problem
- You incorporated feedback into your decisions

If students do the things they need to in order to improve their skills, they will have the evidence to show it. There is still subjectivity in the process, but less than simply looking at the finished product.

There are lots of portfolio platforms that make it possible for students to document their process. Each of these has its pros and cons, but for me, there are a few critical elements:

1. Students should be able to add, resize easily, and crop images.
2. They should be able to include video, audio, text, and other media.
3. All the materials should be in one working document. They should not have to go hunting for previous slides or information.
4. It should be flexible. The teacher should be able to add content through the course.
5. Students should be able to add pages and sections.
6. It should be as user-friendly as possible. Documenting will always feel like work to students. We shouldn't add to that feeling by making it frustrating work.

I use slide-based platforms for these reasons. Document editors have limitations in the content you can add, and placing images is frustrating. In slide-based platforms, it is easy to move elements around, resize them, and place text, arrows, or boxes around things you want to highlight. It's easy to add comments or even your videos and links to a shared digital slide deck.

But the fact is that whatever platform allows you and students to document a project quickly will get the job done. These process portfolios will have a mix of different elements:

- Reflection/visual thinking prompts inserted by the teacher at critical moments in the flow of the projects.
- Periodic progress slides designed by students based on their work over a set period. Teachers should guide them toward specific skills or types of documentation to prompt them to think about what they have been doing more critically.
- Feedback from the teacher on relevant portions of the portfolio related to progress.
- Multimedia content including text, images, video, audio, sketches, diagrams, links, and more.

I don't create a set group of slides and share those at the beginning of the course. Many blank pages encourage students to go ahead and "just fill in" sections of the document before they even have a chance to discuss or reflect on the skills. There are ways to send slides to student portfolios as needed or post slide templates that students copy and paste into their portfolios.

Another significant advantage of adding slides as needed is the flexibility it provides the teacher. Projects will progress differently from one year to the next. Some students may quickly pick projects and get started, while others may struggle with analysis paralysis. Having a set of premade slides makes you feel like you have to "cover the content," which is never a recipe for getting deeply into learning.

Finally, adding slides on the fly allows you to reinvent your content based on the needs of the class you are working with. Working in an unpredictable and fluid environment means that your students may need to be presented with concepts in different ways, with various activities and lessons. If we are providing students the space and freedom to pursue their projects in individual ways, we need to be responsive as well. I have been changing, tweaking, and updating my art and maker classes my entire career — and expect to keep doing so until the day I stop teaching.

9

FROM MAKER CLASS TO MAKER SCHOOL

When students sign up for a maker class, they are expressing an interest or willingness to learn to be creative makers and designers. They may not know exactly what that means, but they are open to it. That also means they already feel like the makerspace is open to *them*, a place where they can fit in or be welcome. What about the many students at your school who do not feel the same sense of welcome in your space?

In 2020, research by AnitaB.org found that in the tech industry, only 28.8% of employees are female. The British firm PWC found several alarming results in their study, including that only 16% of girls had been spoken to about careers in tech compared to 33% of boys, that 83% of boys studied a STEM subject compared to 64% of girls, and that over 25% of female students said they had been put off from a career in tech because it is so male-dominated. While this study was based on UK high schools, these numbers are similar worldwide.

Girls and boys work together in elementary school design classes, engineering activities, and STEM lessons. Girls participate at equal rates as boys and perform as well or better in these activities. In high school, tech and maker classes are often put up against art, music, and drama courses that traditionally appeal to girls. These divisions are clearly social. An 8th or 9th-grade girl who has to choose whether to be the only girl in a tech class or join her friends in a less-preferred elective will have a difficult choice ahead of her.

A robust high school maker curriculum focused on creativity and design can help girls feel more welcome in technology and engineering, if only just for the word "maker" in place of "tech" or "hacker." A maker curriculum that allows students to experience technology and design in ways that are personally engaging to a diverse audience has the potential to break down the barriers that turn girls and students of color off high school tech courses.

The goal is not just to get girls to take more tech electives. The goal is for the tech program to appeal to as wide an audience as possible. A program that appeals to a broad audience will inevitably have a better representation in all demographics.

It's socially tricky for high school students to step out of their comfort zone and try new things. In 8th or 9th grade, they start to think of themselves as "art kids," "band kids," or techies, athletes, readers, etc. The creative student-athlete may not have the time to pursue their main focus and fit a maker class into their schedule. If maker education has the potential to change the lives and minds of students in all pursuits, how can we expose all students to that opportunity?

Makerspace as Library, Not Tech Lab

The best way to expand the reach of maker education throughout the school is to encourage teachers, administrators, and students to think of the makerspace as a library of creative technology.

School libraries are a resource for all members of a school community. Though not everyone may utilize it equally, no student or teacher would feel like they *could* not use the library if they wanted — libraries house resources, books, magazines, entertainment, and reference materials available to the entire school community. The librarian is there to help students and teachers navigate the space, find what they need, and share recommendations and materials based on needs.

Makerspaces have long found homes within libraries because, at their best, they have a shared mission. They both exist to share access to materials, knowledge, opportunity, and expertise. To say that makerspaces are about 3D printing would be like saying that libraries are just about encyclopedias.

Setting up a school makerspace as a resource for all is not easy, but using the concept of a school library can help everyone in the community feel like it is a place they can step into.

More Than Just Dioramas

Building making into the entire school curriculum has to be truly valuable to all constituents. Often, hands-on projects benefit students because they are fun and engaging but are not particularly valuable to the teachers because they are just tacked onto the class content. What do students *learn* through making dioramas? Not much new about the ecosystem or historical event they are meant to be portraying. That learning comes elsewhere; the diorama is just a way to make it more fun.

Another scenario is when the project is valuable for the maker educator but not particularly valuable for the class teacher. Having students 3D print an animal during their marine biology course exposes them to technology, which is great for the tech teacher but doesn't have much value for the science teacher. What do you learn about biology by looking at a plastic animal?

Projects that don't involve learning new skills benefit the classroom teacher because it engages their students in the content without requiring much time. Paper, scissors, and glue are excellent materials. Still, for the maker educator hoping to expose students to the possibilities of the makerspace, these materials don't push students into new realms.

Curriculum-based makerspace projects should be:

- Valuable for the content area teacher in terms of student motivation and engagement as well as helping students learn the content
- Valuable for the students by making learning more exciting and hands-on while exposing them to new skills and possibilities
- Valuable for the maker educator to show students the possibilities of the makerspace to impact any subject that they might be interested in, not just engineering or computer science

Content area classes are how I expose a wider audience to the possibilities of the makerspace. Over 3 years, the percentages of 8th through 12th graders participating in these projects were within a percentage point or two of the overall school population. On average, three times as many students in those grades participated in activities in the makerspace compared to taking yearlong or partial-year electives. In addition, the percentage of female students taking those makerspace courses has risen compared to non-makerspace tech courses like computer science. While the rate of girls in these classes still lags behind the school population, the racial and ethnic diversity of these classes closely matches the percentages across the school.

These numbers represent one school's experience, and the numbers fluctuate from year to year, but what can't be denied is that if you are a student at my school, you will do several hands-on projects in the makerspace even if you never signed up for a class in the space. Those students will walk away with a different vision of technology's role in their lives and interests, even if they lie outside of making and engineering.

Making It Happen

A number of systems need to be in place to make this work, and unfortunately, they are far from the norm. I have visited many schools with a "makerspace" that served only engineering, robotics, and computer science classes. This space is not a makerspace; it's a tech lab. The mission of a tech lab is very different, and so is the audience it serves.

On the other hand, I have visited many schools with a makerspace that is almost entirely unstaffed. Sometimes a full-time teacher receives a stipend (or not) to keep track of supplies and fix things that break. There are almost no structured activities, and occasionally, individual students or a class go in there and mess around on a very surface level, with no opportunity to go deeper.

On the other hand, some schools are so focused on maker education and creative tech that everything they do is built around them. Students lucky enough to get into these private or charter schools have opportunities that their peers can only imagine.

The vast majority of students in the U.S. and other western countries are taking a course of study that looks very similar to what it looked like 20, 50, or even 100 years ago. How can we bring meaningful maker education to these students? One way is to ensure we offer worthwhile maker courses like the one I have outlined in this book. Another is to bring meaningful maker projects into traditional content. Students who enjoy the latter are more likely to sign up for the former.

Maker Educator Schedule

Librarians don't teach five classes a day, and if a school wants to give students access to the makerspace, the maker educator can't either. Schools requiring their maker teacher to cover engineering, robotics, making, or whatever they schedule will not be able to make this system work. That is like turning your librarian into a reading teacher. See how many students get real value from the library then.

What allows me to make this program work is that I teach one scheduled class a day. The rest of my day is set aside for maintaining the makerspace, designing projects with teachers, working with classes on content-area projects, working with individual students or teams that want to use the space for their projects, and keeping an eye on the personal projects (students and staff) that are taking place in the room. It's a full-time job.

I have met incredibly dedicated maker educators teaching four, five, or even more classes in their space and still trying to facilitate content-based projects. They wear themselves out trying to expand their program to students who may not feel confident signing up for a makerspace class. I have watched many of them on social media, sadly taking on other roles or quitting teaching altogether because burnout is a very real thing. Administrators and districts must look at maker educators like librarians and structure their schedules accordingly.

Building a Maker Culture

For those that find themselves in a position to offer meaningful makerspace experiences to an entire school, you must reach out to make it happen. You must be more than welcoming; you must be inviting. You need to do more than leave the door open and wait for teachers and students to come in; you need to reach out and invite people to try something new. So many of the best projects I have done with classroom teachers started with a conversation in the lunchroom after I asked, "So, what are you studying with your students these days?"

There will be early adopters, and there will be teachers that have absolutely no interest in changing what they are doing. Some schools mandate that teachers participate at regular intervals in a makerspace project. To me, this is a recipe for disaster and resentment. Instead of requiring participation, encourage it by showing how valuable it is. When you do projects with your early adopters, showcase the results. Send newsletters about the students' excellent work, create public displays, or have live demos. Other teachers who are open but hesitant will start to reach out to you.

Keep track of which classes and students have taken part in makerspace activities. The goal is not to bring every teacher into the space; it's to bring every *student* in. You can engage every student in the school without ever getting those few teachers who are skeptical or feel like making has no place in their curriculum. If one or two teachers teach all of the 9th-grade history classes, one project can reach every single freshman in the school through just those few teachers.

Offer services to the school in general. Making things that benefit the school beyond the classroom shows that the makerspace is a true resource for everyone:

- Cut vinyl decals with teachers' names for their doors
- Create decorative items with the school's logo
- Make signs for nature trails or cross-country courses
- Fix broken things for facilities or dining staff
- Make stickers for admissions or school clubs

Involving students in these projects makes them even more meaningful, as they see their work around the school and tell others that they did it in the makerspace.

These steps help build a perception of your makerspace as a valuable resource for everyone in the school community. You want students in physics or art classes to ask their teachers if they can get help building some project or experiment in the makerspace. If your theater and athletic directors ask you for help making signs or building sets, you know the makerspace is a community asset.

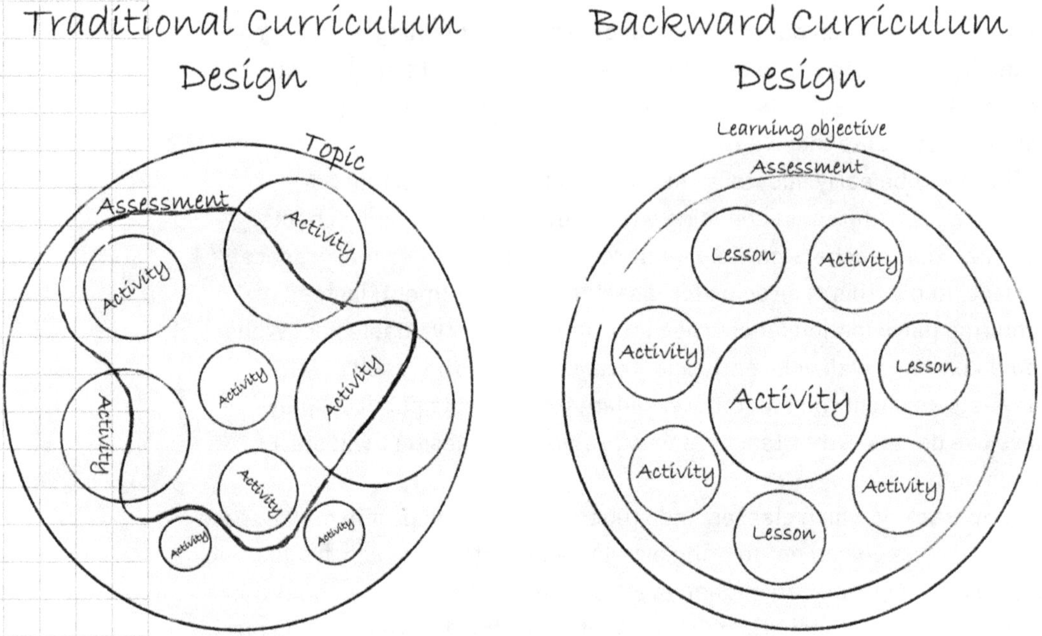

Traditional Curriculum Design

Topic
Assessment
Activity
Activity
Activity
Activity
Activity
Activity
Activity
Activity

Backward Curriculum Design

Learning objective
Assessment
Lesson
Activity
Activity
Activity
Lesson
Activity
Activity
Lesson

Meaningful Projects Through Backward Design

Designing meaningful projects with teachers is the most valuable service you can provide. Meaningful projects go beyond engagement. They are not just to infuse technology into existing curricula; they either act as assessment tools for content already learned or as vehicles for delivering the content in the first place.

I use a process called backward design to make sure projects are meaningful. In the book *Understanding by Design*, Grant Wiggins and Jaye McTighe describe the typical way lessons are made and propose an alternative. Most lesson plans are content based: Teachers pick the book, scientific concept, or math procedure they need to teach, then create a series of lessons, and finish by designing an assessment of what students learned. Through this system, much content covered is not on the assessment. Students constantly ask, "Is this going to be on the test?"

Backward design reverses the traditional process. Teachers first identify what learning objectives they have for their students. Instead of teaching a book, they might identify particular plot mechanisms or writing methods they want students to be able to recognize. Instead of teaching a unit on deserts, they might want students to be able to explain how the process of desertification develops or how certain animals find water. Instead of teaching students differential equations, they may want students to use them to model real-world relationships. The project should start with the statement "I want my students to be able to ..."

The next step in backward design is to create the assessment. If a teacher reaches out to you, they are looking for something hands-on. What object can students design to show they understand the learning objectives? What could they create to put the concept studied into action?

Science and math teachers often want their students to create something functional. It may be an object that demonstrates a concept in physics or fits a mathematical relationship like proportion. Humanities and language-based projects more often include symbols and representations. Students may have to create an object representing a plot theme, a cultural artifact, or a political concept.

Basic Backward Design Process

1. **Identify the learning objective.** What do you want students to be able to do or demonstrate?
2. **Create the assessment that will measure that learning.** What could students build, design, or create to show they understand the concept?
3. **Design the lessons needed to prepare them for the assessment.** What skills and information do students need to complete the project?

Projects don't have to take a lot of time. When teachers are thinking about an object as an assessment, I ask them what time frame they are thinking about to gauge options:

- **Projects for 1–3 class periods:** These replace a test or build a specific object. They are restricted in the process and end result. Students create within a relatively narrow set of objectives.
- **Projects for 1–2 weeks:** These may be used to replace a paper or presentation. Teachers may also use them to introduce or reflect on content learned in class. These projects usually offer more creative freedom, but the objectives are clearly laid out at the beginning, even though the means may not be.
- **Projects for 2+ weeks:** These are designed to deliver content. The creation process is the vehicle that students will use to explore the content. Decisions in the build process will often lead to additional research into the topic, which allows for more creative ideas, which require more research/discussion/processing. The objectives at the project's beginning are fairly broad and conceptual.

This combination of the form of the project and the time available leads to the third part of backward design, which is the steps necessary to create the project.

It may not be possible to create the object the teacher wants in the time available, and this is when to adjust expectations. It's never good to overpromise on a project. Processes like 3D printing take time, depending on the number of printers you have, so time to fabricate needs to be built into the schedule.

There are workarounds, however:

- If teachers can grade digital representations of the objects during fabrication, the project is not dependent on the physical version.
- Some projects can be purely digital creations.
- In some cases, digital designs are created early in the lesson, leaving time to fabricate while traditional lessons take place. The students can return to the makerspace at the end of the unit to assemble the fabricated parts.
- You can shorten a project by limiting the options or pre-making some parts.
- Use simpler design tools that don't require much training. Students can get up and running with Tinkercad in a class block. Google Slides or PowerPoint gives quick entry into vector design. Websites like Makercase.com or Thingiverse.com make creating or finding objects quicker.

Remember that the goal is not to spend much time training art and Spanish students on how to use complicated digital tools. The goal is to expose a wide range of students to the possibilities of the makerspace. For some, this might lead to an interest in digital design, while others may become interested in digital fabrication, and others may get excited about the engineering design process. Some will not become interested in any of it, which is fine. Not every student needs to choose the makerspace over the art studio, or the chemistry lab, or the library. But every student should have a basic understanding of what the makerspace is and how it could help them achieve their needs.

To help me stick to this formula, I created this planning document to help facilitate that process and keep a record of the project for future iterations.

Makerspace Project Design

1 What content, skill, or understanding do you want students to demonstrate?

2 What would students need to do/make to demonstrate this skill or understanding?

Project Design Style 1:
Students execute a design that has been presented to them by the teacher.

- **Duration:** 2–4 class periods (additional time to fabricate completed designs as needed).
- **Student Choice:** Students get creative control over visual details of the finished product.
- **Assessment:** Rubric addresses specific requirements of final product.

Project Design Style 2:
Students are presented with options from which to choose how to respond.

- **Duration:** 4–10 class periods.
- **Student Choice:** Students choose from a menu of options. Creative design control and collaboration.
- **Assessment:** General rubric based on written or oral reflection by student explaining design choice.

Project Design Style 3:
Students lead the design of the project based on understanding of concepts.

- **Duration:** 10+ class periods
- **Student Choice:** Students direct the process with teacher facilitation. Emphasis on group work, consensus, and collaboration.
- **Assessment:** Rubric created with feedback of students as part of the process. Written/oral reflection.

3 Based on the information above, outline the project.

List required elements, project checkpoints, tools needed, activities to do.

THE THREE MODES OF MAKING IN CONTENT AREA PROJECTS

The three modes of making also apply to projects with classroom teachers. In this case, they are less about assessing students' skill development and more about how the projects are designed.

Short projects meant to replace a test will generally be imitation projects. Students will likely have to work from a tutorial or a series of steps to complete the build in the available time. Giving students creative control over the end product is always preferable, which can push the project into modification territory. When students get to create a project while exploring content, they will experience innovation.

Imitation Projects

- Follow a tutorial to create a scientific instrument to measure something — weather devices, inclinometers, ramps, roller coasters, etc.
- Use mathematical relationships to demonstrate a mathematical relationship or structure physically. Measure the center of mass on a flat shape, use the Pythagorean theorem to calculate the hypotenuse of an outdoor patio, and find the center of a circle using rays.
- Create a physical interpretation of an object described in a story.
- Recreate a historical artifact from original sources, drawings, or other representations.

Modification Projects

- Design a flag with symbols representing a culture that you have studied.
- Make paper rockets and experiment with a variety of nose cone and fin designs to test the most effective combinations.
- Create a map that shows the path of migration, refugees, trade, etc., in 3D.
- Engrave a photo of yourself with images of your hopes and dreams. Present these items in another language that you are studying.
- Learn about Frida Kahlo and make laser-engraved selfie frames with autobiographical words and images.

Innovation Projects

- Create an object that represents the transformational journey of the main character in a book, story, or play.
- Create a monument to a historical figure that you have learned about. Create a process of design review and selection. Make the final version at a large scale.
- Create an existentialist board or card game. Don't make it *about* existentialism; make the rules existentialist.
- Use differential equations to code and create a 3D object in Tinkercad Codeblocks. Explain the relationships and how they generated the shapes.

In the next chapter, I dig more deeply into some lesson plans for several of these projects. The goal of these cross-curricular projects is not to directly teach the skills discussed in this book or to load students up with technical information and tasks. The goal is to support their demonstration of content mastery while exposing them to the possibilities of the makerspace.

That doesn't mean that these projects are not an opportunity to extend the invitation to spend more time in the makerspace, whether taking a class or doing other projects there. Building a culture of making at your school means more than just signing kids up for classes; it means making it clear to everyone that, even if they don't take advantage of it, the makerspace is there for the many, not the few.

10

PROJECTS ACROSS THE CURRICULUM

You may become very busy when your fellow teachers learn that you will help them design projects for their class. Over the years, I have found that the teachers that approach me fall into a few distinct categories:

- Teachers who have seen a project on social media, at a conference, or from a fellow teacher and want to do it with their students.
- Teachers who have a clear idea of what they want their kids to make, though they may not know the logistics of how to do it.
- Teachers who have a vague idea for a project and need a lot of help in fleshing it out.
- Teachers who just want a hands-on activity for their kids and are open to anything.

As I mentioned in chapter 9, many teachers may not be interested in doing projects with you. However, others will give it a shot when they see the quality work that students do in well-designed projects. Remember that reaching all the students is more important than reaching all the teachers.

The more significant challenge is redirecting the teacher who is excited to do a project but doesn't know *why* they want to do it. We've all experienced lazy writing and silly word problems. Pointless projects are a waste of everyone's time and energy. The makerspace presents the opportunity to make these projects have a purpose.

In this chapter, I share several lesson plans for makerspace projects created with classroom teachers. Some of them I have done once, others have been repeated over the years. They are examples of how I use backward design to ensure that teachers and students get tangible benefits regarding content knowledge, deeper understanding, and hands-on engagement. Though these projects come from specific content areas, they can be adapted to several different topics.

English Projects

Projects in English and language arts classes tend toward symbolism, imagery, and metaphor. These projects go beyond mere dioramas; they require students to design objects representing big ideas, plot points, and conceptual details to make a statement about something they have read.

TRANSFORMATIONAL JOURNEYS

Learning objective: Students should be able to explain the transformational journey of the main character(s) using symbolism and metaphor.

Product: Students will work in teams of two or three to create an object or series of objects that use imagery, symbols, and text drawn from the book. Students will present their object during a class gallery walk and explain how it represents the significant transformations of the main character(s) through the book or story.

Lessons/activities:

- **Day 1:** After reading the book and participating in class discussions on the topic, students will get quick tutorials in the following areas:
  - Finding and downloading existing 3D models to use as elements from online repositories
  - Designing objects in Tinkercad; importing and remixing models
  - Drawing shapes for laser cutting in Google Slides and adding images for engraving
  - Reviewing available scrap materials
  - Sketching ideas for their creations with the remaining time
- **Day 2:** Students who want to use 3D printing should prep their items during this class to start the printing process. Other students sign up for help in shifts as they work on digital and physical designs.
- **Day 3:** Students complete work in the makerspace and sign up for extra time for outstanding tasks.
- **Day 4:** In their classroom, students work to write up their object descriptions and prepare for their gallery walk. Final objects are executed in the makerspace on 3D printers or other machines.
- **Day 5:** Students present their objects, describing how the symbolism and imagery represent the transformation of the main character(s) through the story.

Assessment: Does the object include a thoughtful and deep representation of the transformation of the main character (thoughts, revelations, significant decisions) or are the images selected surface-level interpretations of that journey (physical appearance, location, demographic information)?

Student project responding to the book *Kabul Girls Soccer Club*

Student project responding to *The Boy Who Harnessed the Wind*

EXISTENTIALIST BOARD GAMES

Learning objective: Students should be able to explain the basic concepts of existentialism after reading *The Stranger* by Albert Camus. They should understand the concepts of free will, creating meaning or purpose for ourselves, and self-determination.

Product: Board games typically use a system of imposed rules as players act within the rules to try and win the game. None of these ideas fit with existentialism. What "rules" would a board game need to represent existentialism? In groups of two or three, design, create, and playtest a board game that uses existential philosophy as the basis for its play.

Lessons/activities:

- **Day 1:** After reading the book and participating in class discussions on the topic, students will get quick tutorials in the following areas:
  - Designing objects in Tinkercad; importing and remixing models
  - Creating visual elements in Google Slides, including cards, board, standees, maps, etc.
  - Sketching ideas for their creations
- **Day 2:** Quick prototype day. Using only paper, cardboard, and other readily accessible materials, students will start to play their initial game idea, refining their ideas and generating a set of rules.
- **Day 3:** Finalize the rules and start designing game elements. Prepare any 3D-printed game elements to be produced at the end of class.
- **Day 4:** Finish designs and fabricate: print boards, cards, and other game elements.
- **Day 5:** Present finished games and share the rules. Describe how they follow existentialist philosophy.

Assessment: Do the game rules use existentialist philosophy as their basis (players create their own goals or paths), or does the game simply inform about existentialist philosophy (trivia questions)?

History Projects

Much of our understanding of history comes from objects, images, and recordings of events and cultures. Symbols embedded in these objects and interpretations of recordings and narratives can lead directly to creating physical things that help show a student's understanding of historical narratives, timelines, and events. History projects tend towards imagery and symbolism but, to a greater extent than English projects, those images and objects are tied to concrete artifacts and events.

UPDATED ARTIFACTS

Learning objectives: Students should be able to identify key symbols and themes represented in the artifacts of an ancient civilization and explain how they are connected to the culture.

Product: After studying an ancient civilization and learning about the artifacts of that civilization, students will design an object that society might produce if it were still around today and had access to digital fabrication tools. What themes, symbols, and significant cultural influences would be in a contemporary object made by the civilization?

Lessons/activities:

- **Before coming to the makerspace:** Students will work individually or in small teams to research various aspects of an ancient culture. The GSPRITE model is an excellent way to organize research. Students should explore a variety of artifacts and how they represent the culture.
- **Day 1:** Students will get quick tutorials in the following areas:
  - Designing objects in Tinkercad; importing and remixing models
  - Creating visual elements in Google Slides for laser cutting/engraving or vinyl cutting
  - Sketching ideas for their creations
- **Day 2:** Students begin work on their design as a group. Students not involved in directly designing objects work on presentation materials about their concept. The makerspace teacher rotates between groups to help with technical questions.
- **Day 3:** Teams that have parts to fabricate will begin making. Job files of multiple parts can be set up to fabricate after students have left the class.
- **Day 4:** If needed, teams get help from the makerspace teacher with fabrication or design issues. Other groups assemble or work on presentation materials.
- **Day 5:** Groups present their designs, using digital representations of 3D models or 2D designs, if needed, while the final pieces are fabricated.

Assessment: This project will result in contemporary objects, such as phone cases, key chains, jewelry, etc., with embedded themes and symbols from the chosen culture. The form of the object should be derived thoughtfully from their research. Does the object simply have an image from the source culture attached to a contemporary object (surface understanding), or does the object show an understanding of the significance of the symbols (deeper understanding)? For instance, a culture that made elaborate gold coins might result in a piggy bank with symbols pulled from those coins or a credit card with designs drawn from the culture's other artifacts.

MINIATURE/VIRTUAL MUSEUM OF HISTORY

Learning objectives: Students should be able to explain an event in history through a curated set of objects and images placed together in an order and arrangement that leads the viewer through the sequence of events.

Product: While this project is similar to a diorama, the concept requires students to consider what they include and how they arrange it. The arrangement and curation of the items will tell the story of the event based on research. Students will print images, create or download 3D objects, or laser cut 2D objects to be placed in a 12×12" lidless box. The miniature museum can be presented as-is, or you can take a 360-degree photo of the inside of the box to create a virtual museum using the website Momento360.com.

Lessons/activities:

- **Before coming to the makerspace:** Students will research their historical event, identifying the cause-and-effect elements.
- **Day 1:**
  - Students will create a four-slide presentation, adjusting the slide size to 8.5×11" landscape to fit printer paper size. If you have a larger printer, this project can be scaled up. Each slide will represent a wall of their museum in the order a viewer would experience it.
  - Students will learn various ways to search for 3D models that might represent artifacts for their exhibit. The 3D model search engine Yeggi.com helps search many sites. The Smithsonian Institution also has many 3D models of historical artifacts.
  - Students will place images, text, graphic elements, and images of 3D models on the "walls" of their exhibit. These images will help students identify the scale models that should be printed.

- **Day 2:** Students will continue to design their museums, focusing on the 3D models and scale. Print jobs can begin when students are confident of their models and sizes. Students can import models into Tinkercad to create pedestals for the objects.
- **Day 3:** Remove the images of 3D models and print the four walls of the exhibit when they are complete. Glue the photos in place on the inside of the 12×12" box, paying attention to the proper order of walls. Place the 3D-printed artifacts in the exhibit.
- **Add-on activity:** Use a 360-degree camera placed in the exhibit's center to take an image. Upload the image to the website Momento360.com to make a VR version of the display. Students can record audio comments for their artifacts and share the exhibit online.

Assessment: Did students include relevant images, text, and artifacts? Did they arrange them in a manner that shows they understand the cause-and-effect relationship between events, using graphic elements and text to help the viewer understand?

World Languages Projects

World language projects can often overlap with English and history projects since they tend toward cultural understanding or representing personal elements in another language. Projects more targeted towards the world language teachers result in objects students can use when speaking. It is easier to speak fluently when you have something tangible you have created and know deeply. These projects become a prop students can use when speaking before others.

PERSONAL FLAG DESIGN

Learning objective: Students should be able to represent themselves, their interests, personality, and experiences through symbols. Students should be able to speak about these symbols in the language they are studying.

Product: Students will design a flag that represents themselves. They can use Google Slides or PowerPoint to create their designs, and one of the following three tools to fabricate the flag:

1. Direct to garment printer to print directly on fabric
2. Laser cutter to cut felt parts to be glued on a backing fabric
3. Vinyl cutter to cut vinyl decals to be layered to create the flag

Activities/lessons:

- **Before coming to the makerspace:** Students should list personal attributes, significant events, and important life experiences. They should identify symbols that can represent these items.
- **Day 1:**
  - Watch the TED Talk by Roman Mars, "Why city flags may be the worst-designed thing you've never noticed." Discuss the video and what makes a well-designed flag.
  - Demonstrate creating shapes in Google Slides/PowerPoint. Discuss basic shapes as well as how to create custom shapes with polyline and curve tools.
  - Discuss the difference between raster and vector and demonstrate how to trace raster images with a polyline.
  - Begin working on designs in an 8.5×11" format.
- **Day 2:** Complete design work. Begin fabricating pieces in vinyl or felt as needed.
- **Day 3:** Print and assemble designs.

Assessment: When students describe the elements of their flag, it will be apparent if they considered form, color, and arrangement to represent themselves. Students with literal shapes without regard to other design considerations will have less to speak about.

BUILD AND NAVIGATE A CITY

Learning objective: Learn place-based vocabulary around the city (school, museum, store, etc.) and navigation language and phrases.

Product: Students will use Tinkercad to create a 3D city that includes a variety of structures, roads, and intersections. Starting at a set location, students will record instructions on navigating to various landmarks using the language they are learning.

Activities/lessons:

- **Day 1:** Students will get an orientation to Tinkercad and learn how to use basic shapes to create various building structures and designs. Students begin to work on their city layouts.
- **Day 2:** Students continue to design their city and include text signage identifying their required structures.
- **Day 3:** Students complete their designs and begin to record instructions

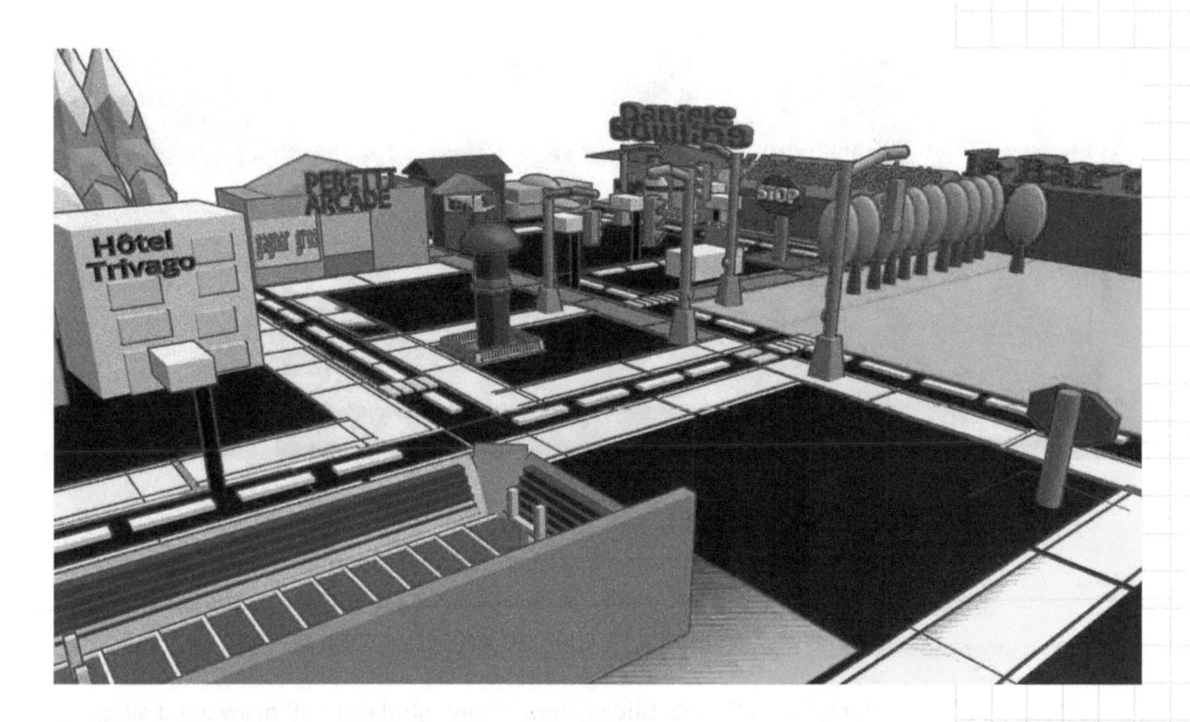

on navigating to particular landmarks using directional language and phrases. They can screen-record movements by placing a character in their 3D design and using the arrow keys.

Assessment: Buildings often have characteristic features, and if students have included these in their models, it will be apparent that they can identify the vocabulary. Moving the character while giving directions will demonstrate that students can understand and use navigational language well.

Science Projects

Science projects usually result in products that do one of two things: model the real world or collect data about the real world. There are many objects that students can create that can be used in experiments and that allow them to visualize some of the hidden aspects of the world and how it works. Successful objects for science should be based on facts and scientific understanding rather than interpretations or opinions, as we have seen in humanities-based projects.

COLORFUL CORAL

Learning objective: Students should be able to identify the individual element of a coral form and the corallite and demonstrate how a variety of coral shapes are created through the growth of coral polyps.

Product: Using Google Slides/PowerPoint, students will draw out a single corallite structure, the hard skeleton of a coral polyp. By copying and pasting this shape and repetitively moving each corallite structure, students will "grow" a coral structure. The resulting design can be laser cut and engraved from acrylic or other materials on a laser cutter or cut out of vinyl with a vinyl cutter.

Activities/lessons:

- **Day 1:** Students will create their basic corallite shape based on their research on coral. These shapes come in limited options with specific features that the students should mimic. They can begin to copy or duplicate their corallite, expanding their design up or outwards to mimic a typical coral growth pattern.
- **Day 2:** Students should complete their coral structure. Students can use paint on their acrylic or other material before it goes in the laser cutter. The shape's outline will be cut, while interior lines should be engraved or marked on the material.
- **Day 3:** Continue and complete fabrication.

Assessment: It should be clear from the design that the student knows what basic corallite structures look like and that the growth pattern results in an existing coral form.

3D CONSTELLATIONS

- **Learning objective:** Students should be able to describe the relative distances of stars from the Earth, their magnitude, and how these factors determine the constellation's appearance from our perspective.

Product: Create an augmented reality experience showing the stars of a constellation across a 3D space. It can also be done with a physical product.

Activities/lessons:

- **Before coming to the makerspace:** Students should select a constellation and identify the stars and their distances from Earth. They should also make a note of their star type and magnitude/size.
- **Day 1:**
  - Using Google Slides or PowerPoint, students will create a .svg diagram of the flat constellation using circles placed on top of a high-quality image of the constellation.
  - Import the .svg file into Tinkercad. Place spheres into the indicated areas on the extruded .svg. Rotate the sphere vertically.
  - Using the distances from the Earth, use the arrow keys to move the spheres deeper into space, using the workplace to judge distances correctly. Measurements will change depending on the constellation.
- **Day 2:** View the constellation from the front with a perspective view. Adjustments must be made to the stars' left/right and up/down positions to put them back into the proper places as seen "from the Earth." Use the original .svg file as a guide. Alter star sizes and colors as needed to represent other details of the constellation as desired.
- **Day 3:** View the constellations in augmented reality. This process can be done directly through Tinkercad on the iPad. Web AR (AR platforms that work on a phone browser with no app required) will make the project even more accessible. AR Studio (web-ar.studio) is a free option for web AR that students can do themselves:
  - Export the Tinkercad design as a .GLTF file.
  - Open a project in AR Studio with a QR code recognition trigger.
  - Upload the 3D file and place it over the trigger QR code. Adjust scale and placement until the constellation fits the QR code.
  - View the QR code and print it out. Scan the code with a phone camera or QR scanner app, and you will be able to move around the constellation in 3D.
  - Adjust the scale and placement until everything looks right.

QR code for a Big Dipper AR model

Assessment: It will be evident through the process of designing their constellation whether students grasp the concept of interstellar distances, 3D space, and how the position of objects is based on specific points of view. Representing scale distances, sizes, and colors will demonstrate an understanding of the various star types.

Math Projects

Math projects usually focus on how math works on real-world objects and situations. So much of the math curriculum is based on abstractions that makerspace-based projects give students a glimpse into how these concepts work in the real world. Since all computer-controlled machines in a typical makerspace use math to direct their movements, it is relatively easy to have them execute precision projects with mathematical applications.

PREDATOR-PREY DIFFERENTIAL EQUATION

Learning outcomes: Students will be able to describe how differential equations can model systems in the real world.

Product: Students will manipulate a code-based 3D object showing the prey population on its front face and predators on its top face. They can determine if the variables lead to a stable predator-prey relationship by altering the code.

Activities/lessons:

- **Before coming to the makerspace:** Students will learn about the longest-running predator-prey study: the wolves and moose of Isle Royale. Various YouTube videos describe the research and recent issues encountered due to global warming.
- **Day 1:**
  - Introduce students to the Tinkercad Codeblocks design found at bit.ly/tinkercadpredprey. This model uses a set of differential equations to show how a predator-prey relationship reaches stability over time.
  - Students break the system's stability by changing the variable for prey to a number double the original. Tinkercad cannot handle a divide-by-zero scenario, so the resulting model is broken.
  - Have students try and bring the system back into balance by predicting how the other variables will affect it. They should make predictions, test them, and report the results.
- **Day 2:** Have students collect and record their findings as they explore variables such as prey growth rate, predator mortality, carrying capacity, etc. If they find solutions, they should try and explain what the variable changes mean in real life (lowering prey growth rate might mean removing infant moose or using contraceptives on the population).

Assessment: If students can describe in non-mathematical language how the changes in variables represent actual actions scientists might take to stabilize an ecosystem, it will suggest they understand how the interactions of the parts of the system can be modeled and predicted mathematically.

FIND THE CENTER OF MASS

Learning objectives: Students will show they can calculate the center of mass of a complex object by breaking it into simpler shapes, each of which can be calculated, and then using those results to find the overall center of mass.

Product: Students will create a laser-cut copy of a complex shape made from several simple shapes with a hole placed at their calculated center of mass. When a string with a knot is put through this hole, the object should hang flat when suspended.

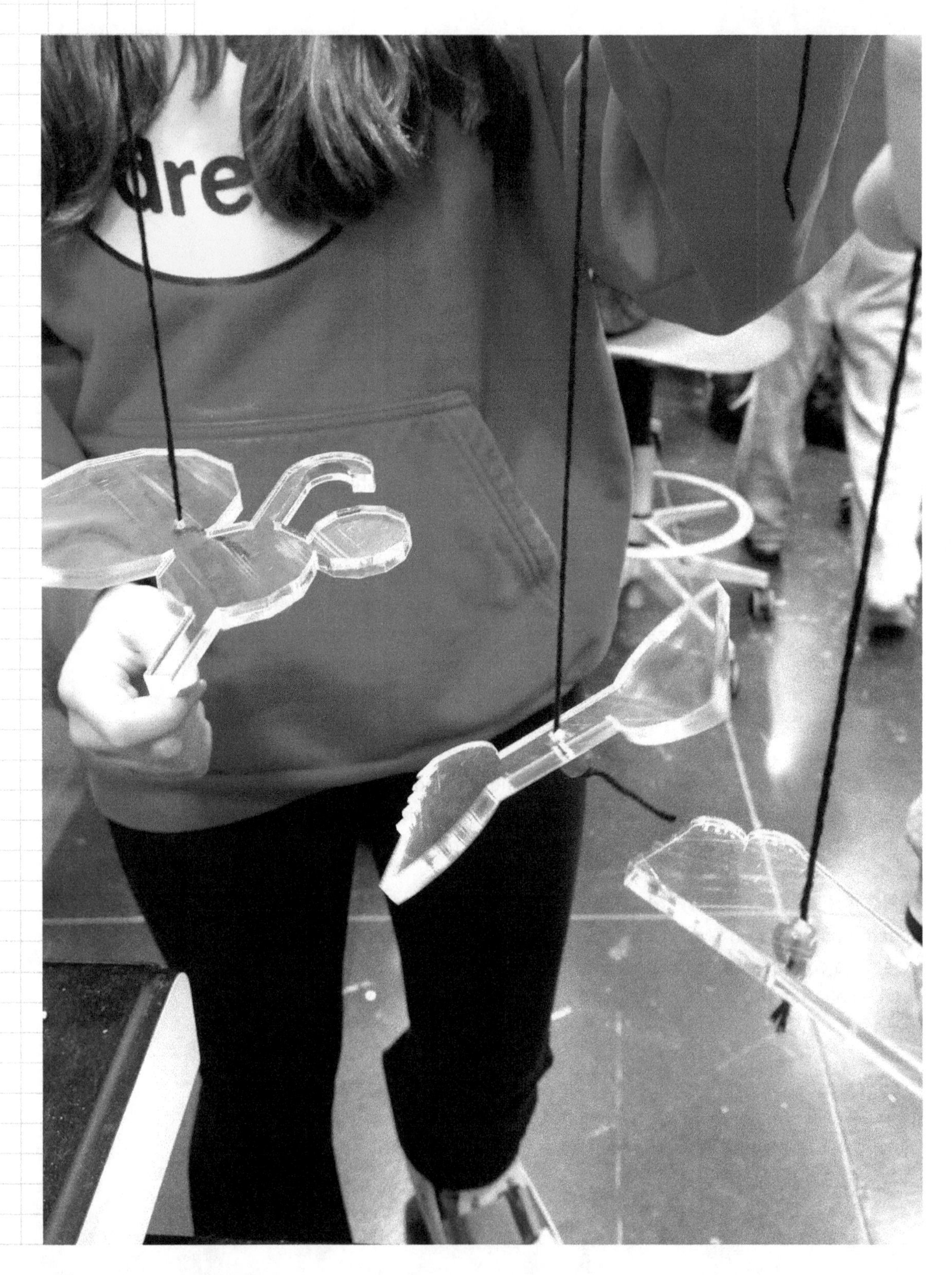

Activities/lessons:

- **Day 1:** Students will make a complex shape out of simple shapes: squares, rectangles, circles, etc. They will use Google Slides/ PowerPoint to create a compound shape representing a real object, such as a pine tree made from triangles and a rectangle or a car made from rectangles and circles.
- **Day 2:** Students will calculate the overall center of mass and place a 2mm circle on that spot.
- **Day 3:** The compound shape and the 2mm hole will be cut out of acrylic on a laser cutter. When a string with a knot tied in it is placed through the hole, the shape should hang flat on the knot.

Assessment: For this project, the final product's success is the assessment. Students with objects that do not hang flat should go back to determine why their calculations are wrong.

Arts Projects

While I have done art class projects in the makerspace, it is much more common for individual students to approach me to create some aspect of an art project that they have planned in their art class. In this way, the makerspace becomes an extension of the art studio, and vice versa.

Projects that do take place in the makerspace often focus on the use of digital fabrication tools to be used to supplement or augment the standard tools of the art studio.

VINYL-CUT SCREEN PRINTING

Learning objectives: Students will learn the process of screen printing and how they can use digitized drawings to create custom vinyl stencils.

Product: Students will make drawings that address the artistic prompt presented. Students should use black ink shapes and lines to create a graphic image. Once the drawings are digitized and vectorized, they can be cut out of vinyl and adhered to silk screens for screenprinting.

Activities/lessons:

- **Before coming to the makerspace:** Students will complete their drawings in black marker, filling all areas they want to have color in black. These drawings do not need to be large.

- **Day 1:**
  - Students will scan their drawings using a phone or scanner. The app Adobe Capture can turn phone camera images directly into vector files. Use a program like Inkscape or Corel Vector to vectorize drawings if needed.
  - Students begin cutting their designs out of vinyl sheets sized slightly smaller than the screenprinting frame.
  - Show students how to weed their vinyl design.
- **Day 2:** Complete the vinyl cutting and weeding. Apply the vinyl decals using transfer tape to the screenprinting frames and use tape to seal the edges.
- **Day 3:** Begin screenprinting.

Assessment: The assessment for this project will be based on the initial prompt for the drawing assignment.

DIY AMPLIFIED PERCUSSION INSTRUMENTS

Learning objective: Students will identify materials to make interesting percussion sounds, then use these materials and electronic components to make amplified instruments.

Product: Using inexpensive electronic components (contact microphones, 3.5mm aux cables, powered speakers) along with found materials, students make amplified percussion instruments.

Activities/lessons:

- **Day 1:** Students will explore a variety of found objects and materials and explore ways to make sounds through percussion. They will combine various materials to create a percussion instrument.
- **Day 2:** Using contact microphones that turn surface vibrations into electrical signals, students will wire the microphones to their instruments. Multiple microphones can be used to take advantage of sounds from various locations or percussion methods. Connect the microphones to a 3.5mm aux cable using wire or conductive tape.
- **Day 3:** Connect the aux cable to powered speakers or a small amplifier and test the instruments. Adjust the microphone or wiring as needed to improve the sounds. Make music!

Assessment: Students will be assessed on their creative use of materials and exploration of different sounds produced. Using a trash can or other drum-like item is a simple application of this process. Exploring non-traditional materials to make unexpected sounds shows engagement with the exploratory nature of this project.

11

MORE CLASSROOM ROUTINES & ACTIVITIES

A makerspace class is an active space. In most ways, it is the reverse of a traditional classroom, which usually includes significant stretches of instruction and short periods of application. Most students find this environment less engaging, but even a makerspace needs some routines. If the teacher does not provide the structure, the students will, and the lessons learned will be haphazard and likely not very deep.

The activities and routines in this chapter are like navigation aids in the wilderness. They don't take away the wild nature of the environment, but they give students tools to help them get to where they want to go.

Safety and Basic Use of the 3D Printer

Safety: There are relatively few safety issues related to 3D printers. I ensure students know where the nozzle is, as this is the part that will get to 200°C or more. Touching this, or the filament that has just come out of it and has not yet cooled, is a bad idea.

Damaging the tool itself is more likely than injuring yourself or others. Wasting material is another area that we discuss as a part of basic use.

- **Evaluating models for printability.** There are thousands, probably millions, of 3D printable files available for free online. Many of them are actually unprintable and will lead to waste or damage to the printer. Checking how many times people have posted actual "makes" of the file will help you know whether it's worth printing.

- **Adhesion.** Ensuring your print sticks well to the build plate is one of the most important steps in printing an object. Objects that come detached from the build plate start to slide around as the nozzle moves, resulting in a mess of filament strands and, sometimes, damaging the printer in the process. There are ways to add a wider base to a print to make it more likely to stick throughout the process.

- **Supports.** 3D printers build objects layer by layer from the bottom up. So designs must include supports (to be cut away after printing) for features or parts that overhang or stick out from the print's main body, such as a hand or arm on a character figure. Otherwise, the printer will start placing filament in midair, attached to nothing, leading to failed or detached prints. Unsupported parts lead to a lot of wasted and discarded plastic.

- **Orientation and scaling.** Learning to look at how the model will print and how changes in orientation and scale affect adhesion and support is important to know. Simply laying a model on its back or side may reduce the necessary supports or increase the size of the base layer.

- **Checking temperatures.** Students should know where to look on the roll of filament to see what the printing temperature range should be and where on the printer to see what the actual temperatures are. Printing at the wrong temperatures can definitely damage the printer.

- **Watching the first layer.** Most 3D printer failures occur during the first layer of the print. Showing students what a good first layer looks like and getting them in the habit of ensuring their print job starts well will save many failed prints.

That's all we talk about with 3D printers. I spend about 30 minutes showing some images and video clips related to these critical elements. We look at some models and decide if they are good to print and how we would do it. Then they find a model that meets these criteria and load it on the printer slicer. We walk through how it should be placed, oriented, and supported. Any student who has gone through these steps with me is welcome to load models independently from then on.

Are they experts? Definitely not. Will there be failed prints? Almost certainly. But each of these failures is an opportunity to relearn and reinforce these basic steps. Using a tool hands-on, in a basic and safe way, will be the best avenue to truly learn how the tool works.

Curating Imitation Projects

In order to combat analysis paralysis when students are selecting their first imitation project, I give them a curated list of projects. This list is drawn from sites like Instructables, YouTube, Printables, and more. So that each student can find something that appeals to them, I select projects that range from functional to decorative, high tech to low, simple to complex.

To create my curated list of projects, I ask myself the following questions:

1. **Is the result potentially valuable to the student?** Many projects result in cool and exciting things for me but not for a teenage student. Over the years, I have discovered that I have been wrong about some projects, and removed them from the list.

2. **Is the project doable by a new maker in the time allowed?** I usually give 1–2 weeks of classes to complete the project, which translates to 4–8 hours of class working time. These projects should be short and sweet, focused on building foundational skills and knowledge.

3. **Does the project have clear steps, instructions, and visuals?** A fast and cool-looking project with poorly written instructions is a recipe for frustration. Make sure to look over the instructions from the viewpoint of a new maker.

4. **Does the project assume any prior skill experience?** Project creators often include steps assuming the maker already has basic skills or knowledge. Sometimes it's possible to scaffold this or modify the project, so students don't get hung up on details that are not fully explained. If you can't clarify, look for alternative options to create the same object.

5. **Does the project offer a valuable experience that students can build upon in future projects?** The goal of the imitation project, at least from the teacher's perspective, is to give students experience with tools and processes that can grow into ever more complex projects.

6. **Do my curated projects represent a wide range of interests?** A good maker program should appeal to as wide a variety of student interests as possible. While 3D printing is fun and intriguing for many students, it's not what every student wants to spend their valuable makerspace time on. Students should be learning creative process skills as well as technical skills, so we want them to be motivated to engage fully with the project. If you teach specific tools, offer as many end products as possible, so students get personally invested in the results. If every student walks away with a 3D-printed fidget spinner, you've missed an opportunity to show how students can use a 3D printer to solve many different problems.

The Marshmallow Challenge

The Marshmallow Challenge has its own website, Marshmallowchallenge.com. It involves giving small teams of three or four the following supplies:

- 20 sticks of dried spaghetti
- 1yd of tape

- 1yd of string
- 1 marshmallow
- 18 minutes

The challenge is to build the tallest freestanding structure capable of supporting the weight of the marshmallow at the top.

Many team and collaboration-based lessons can be learned and discussed through this activity; however, for the creative design student, lessons about iteration and identifying the most significant challenge early are the most directly relatable.

After doing the challenge with students, I always have them watch and discuss Tom Wujec's TED talk, "Build a tower, build a team." In this short video, he discusses the importance of putting the marshmallow on top at the beginning and why kindergartners do so much better at the challenge than business school graduates.

The Deep Dive

> ***Enlightened trial and error succeeds over the planning of the lone genius.*** —The Deep Dive: One company's secret weapon for innovation," *ABC News Nightline*, July 13, 1999

A well-known video that I watch with my students is an *ABC News Nightline* story about the design firm IDEO. This video, titled *The Deep Dive*, can be found on YouTube in three parts and shows IDEO's design process as they are challenged to rethink and redesign a shopping cart. I show this video to students who are about to embark as teams to design something innovative for a client.

I highlight several elements:

1. IDEO is not an expert in making any particular thing. They are experts at applying their process to all sorts of things. For students who are not confident in their creative or technical skills, this is a good reminder that having teams with wide-ranging experiences and interests can be a plus.

2. They get out into the world and develop empathy for their clients and those affected by the design. They talk to parents with kids of all ages, employees of the stores, casual shoppers, and professional shoppers about the carts. They do their own observations and research.

3. They use divergent thinking to generate all sorts of ideas, and only allow supportive comments. They encourage wild ideas that don't seem reasonable or feasible because of the way they help drive innovation forward.

4. The team works within an environment of "focused chaos." The process looks messy from the outside, but the activities they take part in are designed to involve everyone, generate the best ideas, get feedback on them, and ultimately come up with a successful solution.

5. They make rapid prototypes out of simple materials. They make mockups of their big ideas in wood, PVC, plastic, and more.

6. They move from a divergent to a convergent process to choose the key design requirements. The team leaders make final decisions and the team acts on them.

7. They make the final idea a reality by fabricating to as high a level as they can.

8. They take the final design out into the world, share it with people, and continue to get feedback on how it could still be improved.

This video shows how a group of motivated individuals, working freely within a purposeful creative process, can imagine, assess, and ultimately create better designs than a single individual, no matter how brilliant.

Making and Tinkering Skills Identification

The Exploratorium is a public museum in San Francisco that explores science, art, and technology in hands-on ways. In 2017, educators from the museum studied how their after-school making activities resulted in various maker skills that could support traditional educational skills. They published a professional development tool they call "Learning Dimensions of Making and Tinkering." You can find it on their website at Exploratorium.edu/tinkering/our-work/learning-dimensions-making-and-tinkering.

Their framework identifies skills in several key areas of the creative process. These include:

- Initiative and intentionality
- Problem solving and critical thinking

- Conceptual understanding
- Creativity and self-expression
- Social and emotional engagement

Many of the skills they identify, such as "setting one's own goals," are skills that I highlight as part of my maker class curriculum. Others, such as "making observations and asking questions," are skills I highlight in rubrics to identify when students apply skills.

As helpful as this framework is for me, it's also valuable for my students to take this umbrella view of the creative process. In this activity, we watch a YouTube video that features a maker's process, and students identify the moments when the maker uses the various skills on the list. I like to use a project video showing a process, not just a finished product, that involves the maker engaging in problem solving and experimentation to arrive at their completed project. YouTubers Evan and Katelyn Heling are great at showing their iterations and experiments. In addition, Ben Uyeda's YouTube channel *Homemade Modern* has straightforward narratives where he often shows different ways you can make his designs using a variety of tools and techniques.

Once we have watched the video and students have made notes on their observations, we catalog them together. Since this is the same process teachers use when documenting their observations of students, it gives them a glimpse into examining the process rather than the product.

Diagramming the Design Process

The International Baccalaureate (IB) program is a curriculum that traditional schools can offer their students in addition to their typical curriculum. The IB program has a heavy emphasis on critical thinking and engagement. The IB Middle Years Program (MYP) has a design cycle diagram that visualizes what designing and building a complex project looks like.

Similar to the making and tinkering skills identification activity, I have students watch a creative process video and take notes when they see certain design moves taking place. The MYP Design Cycle is more concrete than the "Learning Dimensions of Making and Tinkering" tool. It includes items like "demonstrating technical skills" and "constructing a logical plan." These are task-based skills rather than creative moves.

Another nice thing about the design cycle diagram is that the arrows point every which way through the process. The creative process is never as cyclical as many descriptions would lead us to believe. The reality is that results and feedback will push us into different phases of the cycle as needed. Viewing the overall cycle should remind us to avoid getting stuck in one area of the process, but we should not seek to go through the process in sequence.

My favorite videos to illustrate this experience are Adam Savage's *One Day Build* videos. These range from 20 to 80 minutes and document an array of projects, from functional to decorative. The beauty of these videos is that they show Adam's creation process and problem solving and include tests, experiments, solutions, mistakes, and successes on the way to the final product. These videos show a lot of machine tools and techniques that students will likely not use in their projects. Adam's descriptions benefit from his years as a TV presenter, and learning to get the gist of something you don't have experience with is also a good thing to practice.

Visible Thinking Routines and Reflections

Goal Setting: A good goal needs to be challenging, achievable, and measurable. Using show and tell methods, describe how your Imitation project meets these standards.

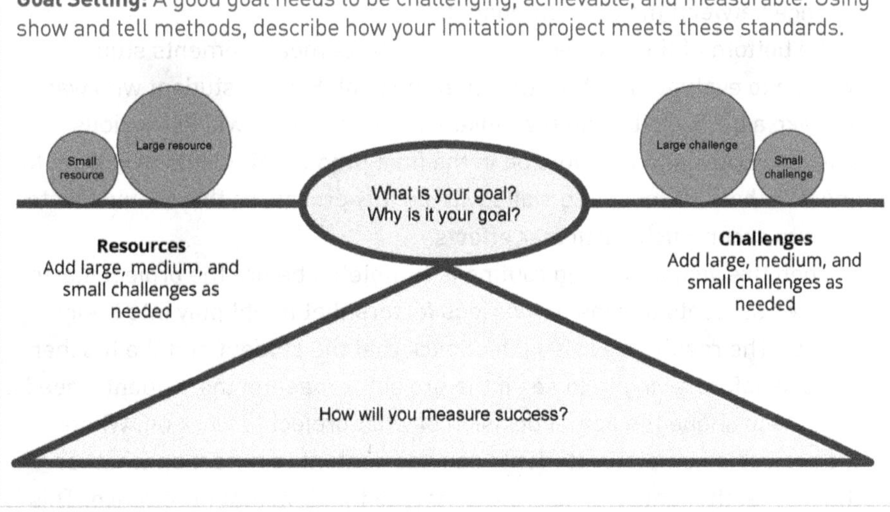

SETTING GOALS

As described earlier, I use the factors of *challenging, achievable, and measurable* when I have students write goals. These three elements are interconnected: the challenges to overcome, the resources you have available, and how to measure whether you're successful. If students identify many challenges and have only a few factors that make the goal achievable, they should reconsider their goal. If they have a small challenge and a large amount of background skills and experience, I encourage them to take on a more challenging task. If they cannot find concrete ways to measure the success of their project, they should get more specific in how they formulate their goal and expected outcomes.

The visual thinking routine I use for this focuses on balance. The fulcrum of the diagram is where the student writes the goal. It is broken into two components: the purpose and how the student will accomplish it. For instance, a student who wants to design an object in Tinkercad may wish to have that object to give as a gift, or they may want to improve their 3D modeling skills. The details of the goal will determine how to measure success in their final section.

The two sides of the balance beam represent things that make the goal achievable (tools, experience, resources) and things that make the goal challenging. These items are placed in circles, and students adjust the circles' size to visualize the item's relative significance. In this way, a single considerable challenge might be balanced out by a lot of little resources, or vice versa, but a quick visual inspection will show whether there is a balance between the two components.

The bottom of the diagram lists the concrete measurements students will use to evaluate whether they met the goal. For the student who wants to make a gift by following the Tinkercad tutorial, they will list various objective criteria that should be in the final product. Students who want to improve their 3D modeling skills will identify processes they would like to learn to accomplish particular effects.

When this visual thinking routine is completed before the project starts, it forces students to consider various factors that might play into their project. The result gives visual feedback that the student and the teacher can use before diving in to see if the project is meeting the student's needs. It can help shape the actual decision of what project to work on. When the project is complete, students can add content to reflect on whether their initial thoughts on the project matched their actual experience. This reflection becomes more meaningful since they already put thought into the front end and executed the project more intentionally.

SOLVING PROBLEMS STRATEGICALLY

Early in the process, problems will likely be more about the overall size, form, materials, and tools. Later on, issues will likely be more technical or specific. The goal is to have students write a statement about their current

What is your current problem or challenge to solve? Use details	
1. I assume that... • Assumption 1 • Assumption 2 • ...	2. It might also be the case that... • Alternative 1 • Alternative 2 • ...

If I use these strategies, it could help me by... (When you complete these, rank them in order to try.)

Research:
Tinkering:
Testing:

problem with as many concrete details as possible. A problem statement like "Should this be made out of plastic or wood?" will lead to a more precise next step than "What should it be made of?"

Deciding to limit ourselves to plastic or wood for the final product can be helpful, but it's important to know why. If a student makes this choice for a specific reason — perhaps they have experience working with those materials — then they have made a choice based on gained knowledge. If they chose them because they didn't know metal and fabric were options, they are basing their decision on false assumptions. It's vital to make a problem statement based on as few assumptions as possible.

For research, they should use the skills they practiced in imitation projects and evaluate how good the source is and whether they have the skills to follow the video. Often students regurgitate big words and concepts without understanding because they think that is what their teacher wants. Having students write or describe what they have researched in their own words will help you determine if they understand the concepts or not.

In science experiments, we use the terms *independent* and *dependent* variables. A good experiment should have only one independent variable, such as the prototype's material. The changes that come about from testing the independent variable are the dependent variables, of which there may be several. Examining how the dependent variables change will give concrete feedback on the independent variable. A/B testing is a way to explore a cause-and-effect relationship.

With A/B testing, students will create prototypes with both materials, or some other quality being the only difference. Students can test them to see which will work better for their project. Determining this will be based on the needs of the project. Are they looking for the strongest material? Are they looking for the material that people like the best? They may want to know what is easier and quicker to produce. Understanding the results of changing certain elements of a project can help a student meet their goals for the project as a whole.

INTERVIEW TENNIS

When students work with clients, they need to develop empathy. Often student interviews go something like this:

- They do a little research or brainstorming
- They write down a list of questions based on that research
- They ask the question in order and take notes

That's it. Everything that comes out of the interview is based on things they already thought were important before they even started the conversation. Great interviews are based on follow-up questions that arise through active listening to responses.

In this activity, we watch a recorded interview with an expert interviewer. I use a sit-down-style interview, not a news broadcast with interview portions. It's best if it is relatively unedited. Charlie Rose's interview with Jay Z is a good example that you can find on YouTube. While we watch the interview, I have students diagram it as a tennis match, pausing periodically to let them take notes. The interviewer's side, the server, has boxes for questions and follow-up questions, and students can add notes to the boxes about the questions. The interviewee's side has responses and significant responses.

Students should see that a significant response should always have a follow-up question. Suppose you are interviewing someone and hit upon something that elicits a significant answer. Significant answers offer information about the interviewee and often are longer or more thoughtful than typical responses. When a student hears a significant answer, their time is better spent following up on that topic to gain more insight than moving on to another question and only skimming the surface. My students prepare 5–10 questions when we do a client interview, but I make it clear that if we only get to three of them because we have so many exciting follow-up questions, then our interview has been a success.

Interviewer -
How to use this slide
Interviewee -
Question
Question
Question
Follow-up question
Follow-up question
Follow-up question
Interview Tennis
Big Response
Big Response
Big Response
Little response
Little response
Little response

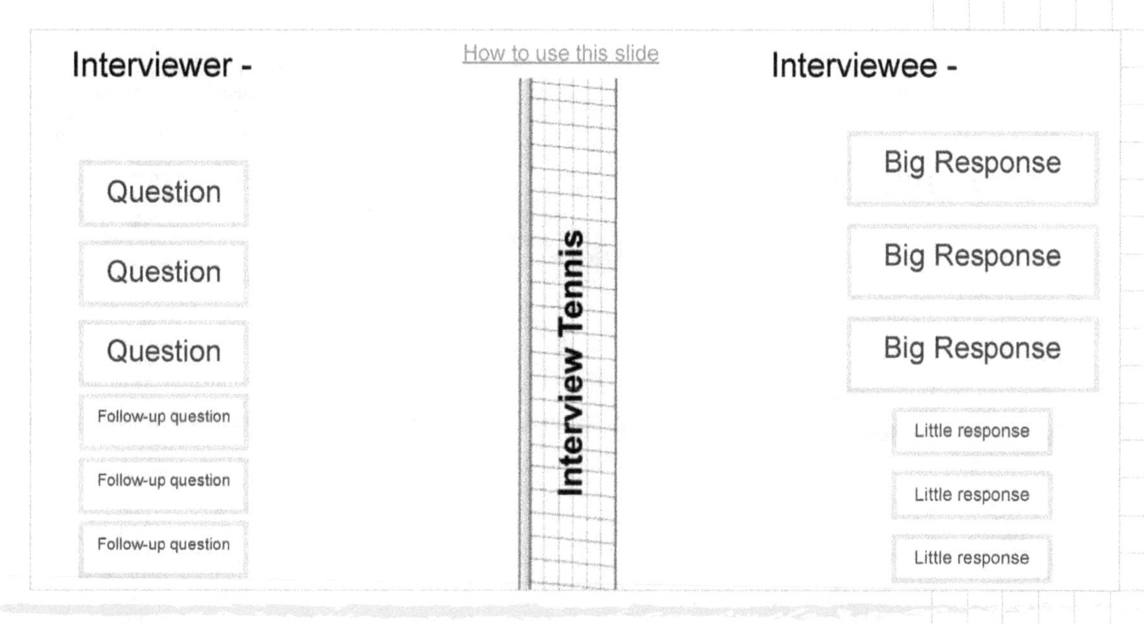

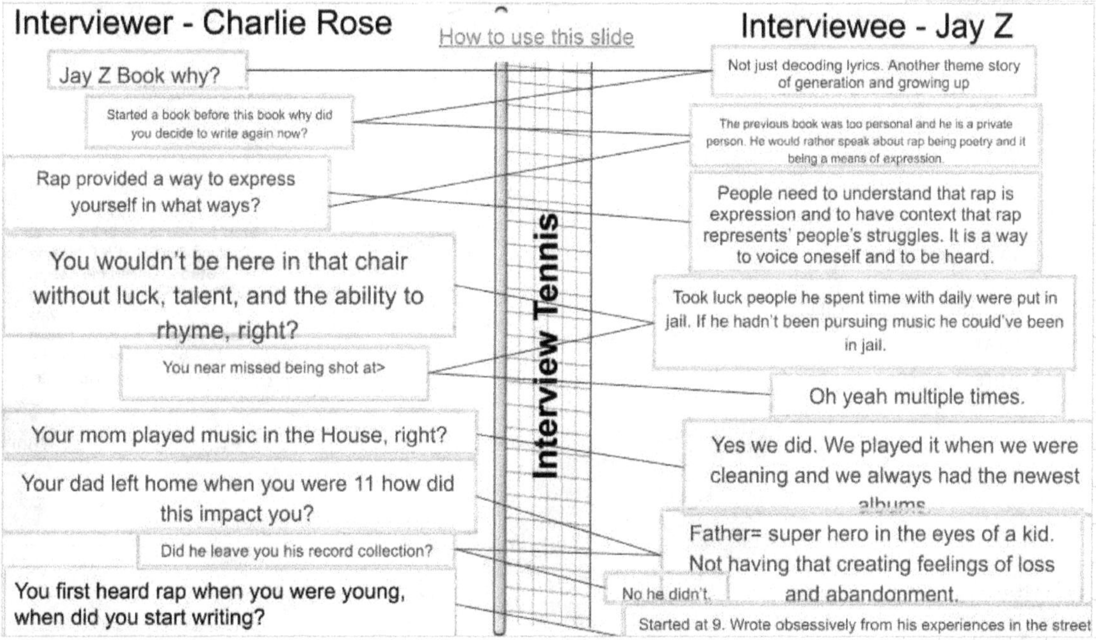

Interviewer - Charlie Rose
How to use this slide
Interviewee - Jay Z
Jay Z Book why?
Not just decoding lyrics. Another theme story of generation and growing up
Started a book before this book why did you decide to write again now?
The previous book was too personal and he is a private person. He would rather speak about rap being poetry and it being a means of expression.
Rap provided a way to express yourself in what ways?
People need to understand that rap is expression and to have context that rap represents' people's struggles. It is a way to voice oneself and to be heard.
You wouldn't be here in that chair without luck, talent, and the ability to rhyme, right?
Took luck people he spent time with daily were put in jail. If he hadn't been pursuing music he could've been in jail.
You near missed being shot at>
Oh yeah multiple times.
Your mom played music in the House, right?
Yes we did. We played it when we were cleaning and we always had the newest albums
Your dad left home when you were 11 how did this impact you?
Father= super hero in the eyes of a kid. Not having that creating feelings of loss and abandonment.
Did he leave you his record collection?
No he didn't.
You first heard rap when you were young, when did you start writing?
Started at 9. Wrote obsessively from his experiences in the street
Interview Tennis

BALANCING IDEAS WITH REQUIREMENTS

When working with clients, makers need to meet specific requirements. These become additional constraints that students must be aware of, but they are targets instead of boundaries. A constraint that is a boundary is self-reinforcing. If students make a phone holder that doesn't fit the phone,

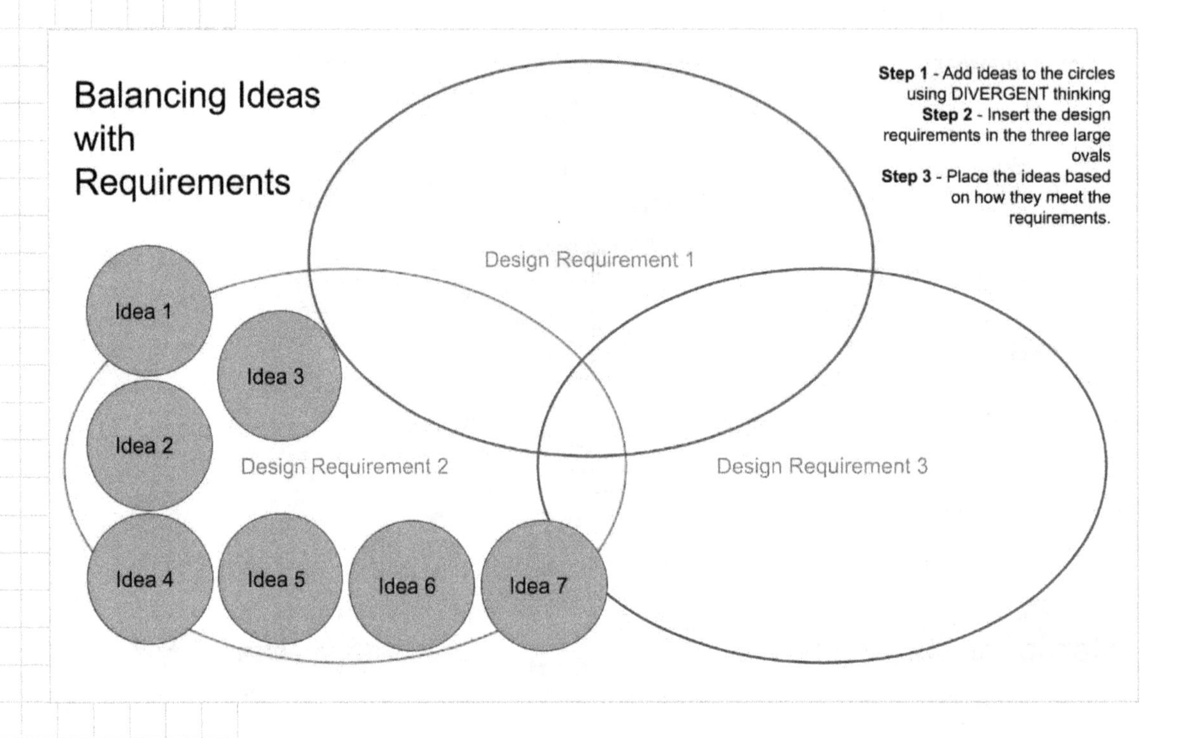
Balancing Ideas with Requirements
Step 1 - Add ideas to the circles using DIVERGENT thinking
Step 2 - Insert the design requirements in the three large ovals
Step 3 - Place the ideas based on how they meet the requirements.
Design Requirement 1
Design Requirement 2
Design Requirement 3
Idea 1
Idea 2
Idea 3
Idea 4
Idea 5
Idea 6
Idea 7

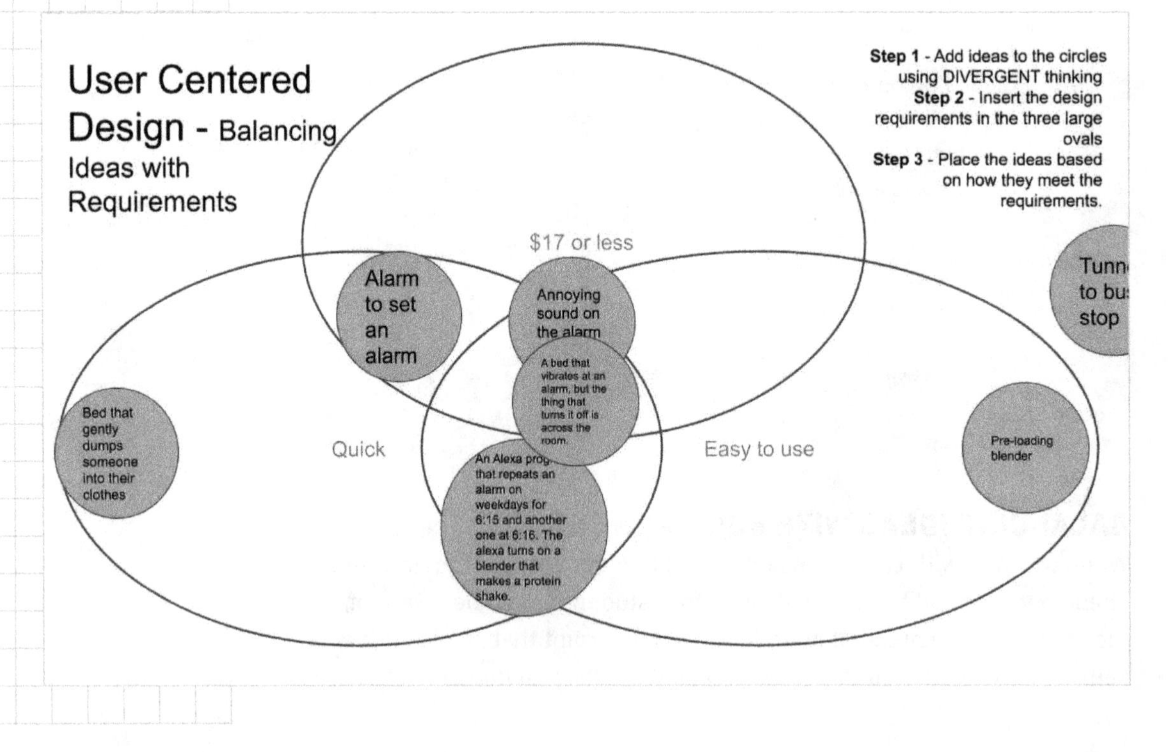
User Centered Design - Balancing Ideas with Requirements
Step 1 - Add ideas to the circles using DIVERGENT thinking
Step 2 - Insert the design requirements in the three large ovals
Step 3 - Place the ideas based on how they meet the requirements.
$17 or less
Quick
Easy to use
Alarm to set an alarm
Annoying sound on the alarm
A bed that vibrates at an alarm, but the thing that turns it off is across the room.
An Alexa prog
that repeats an alarm on weekdays for 6:15 and another one at 6:16. The alexa turns on a blender that makes a protein shake.
Bed that gently dumps someone into their clothes
Pre-loading blender

they must redesign it if they want the project to succeed. A target constraint can be forgotten or ignored if it isn't dealt with purposefully.

To ensure these requirements are front and center, you must first identify what they are. This decision should be based on what students have learned in their client interview, research into existing designs, capabilities of the makerspace, etc. Should the design be simple to use? Should it be aesthetically pleasing? What about fully or partially 3D printing? Many factors may be worthwhile, but what are the actual requirements? There should only be a few.

Using the "Balancing Ideas with Requirements" diagram, students should list as many ideas as possible, using divergent thinking skills to generate many options. Once the project requirements have been placed in the Venn diagram, they will move their ideas around based on how they fit. An inexpensive and 3D-printable project that is not simple to use will be placed between the first two circles but not in the center of all three. The goal is to critically evaluate many ideas against these target requirements.

Students should do this activity before and after the project. Did the initial idea hit the targets? Is there still work to do to get it closer to the center of the diagram? The goal is to make something that works for the clients, not just something that works.

THE STRATEGIC PLAN

I do not assign the strategic plan routine until very late in the innovation project, when students really have a lot of knowledge and understanding about their project. Although they may have been working for weeks on prototypes, solving challenges, and testing ideas, these activities are really a form of planning by doing.

The scientist and philosopher Michael Polanyi describes the difference between "explicit knowledge" and "tacit knowledge." Explicit knowledge is what we can learn from manuals, websites, and documentation through research. Tacit knowledge is what we learn through practice, muscle memory, and direct experience. When students have started to develop tacit knowledge about their project, they feel like they have a grip on it and are ready to dive in. When they reach this phase, they are ready to create the strategic plan.

Because it is so hard to put that into words, I again use visual means to represent how they see their project. At the top of the strategic plan diagram, they give their project a descriptive title. In addition to the title, they should describe their project based on their original goal.

Below that they should have three to four main features of the project.

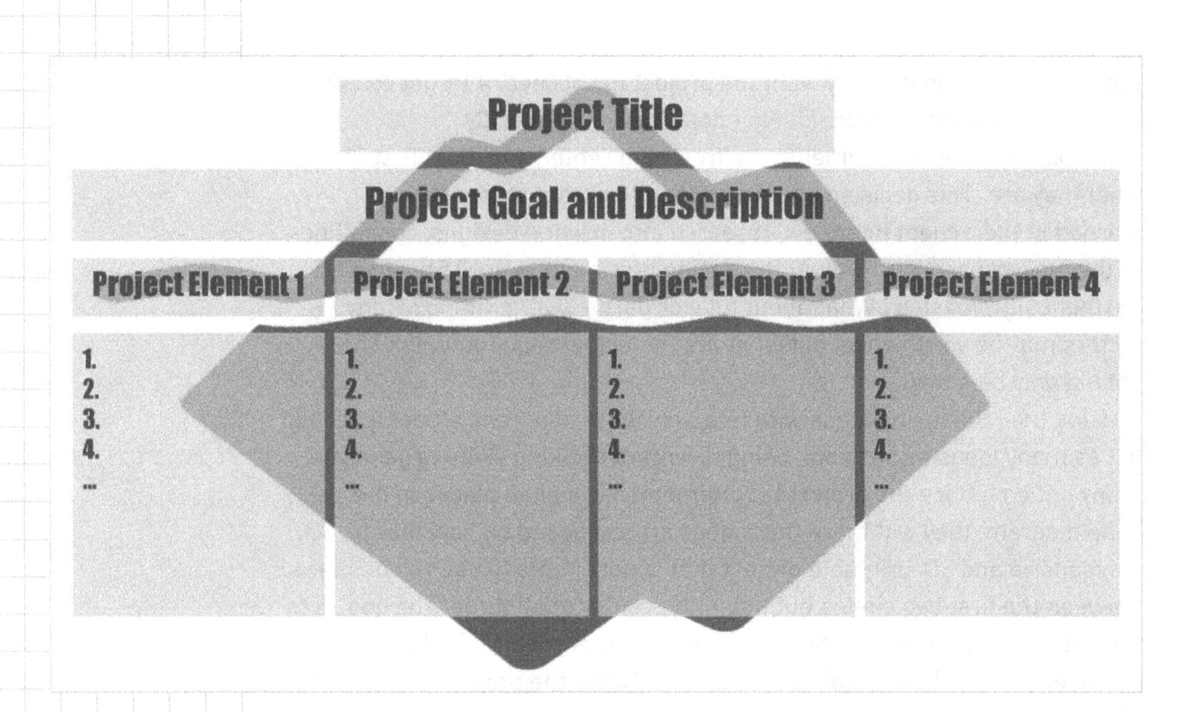

These could be:

- Tools or processes learned to complete the project
- Prototypes or versions of the project
- Major challenges that needed to be solved
- Data or tests that needed to be done to prove the design was viable

It should be up to the student to think this through. This is their project, so they should be able to describe it in their terms.

Under these sections — what Adam Savage would call the Medium Chunks in his *Every Tool's a Hammer* book — they should fill in the details of how they learned the tool, made the prototype, solved the problem, tested their idea, etc. Some of these items may be work to do in the future, and they can be color coded to highlight that the project is still developing.

In the last stage of the project, when students are working to accomplish as much as they can before the end of the term, they can use this plan to check off their steps. When the work is done and they display it for the school to see, this plan should be printed and displayed with it, highlighting all the work that went on below the surface.

WEEKLY PROGRESS/PLAN DOCUMENTATION

There comes a point in a maker course when the students have done all the reflections, have a plan, know the biggest challenge, and must execute their plan. At this point, additional reflections and visible thinking routines are unnecessary speed bumps as students work to turn their ideas into reality. There is still a need to get a look at how the students are thinking about their project and the decisions they are making. This may not be possible through casual observation, especially if you are helping a few students with challenging parts of their project while others appear to be plugging away.

The weekly progress documentation slide is designed to get as much information as possible without taking much of their valuable time. You can quickly grasp how they are tackling their project from a few words, pictures, and key concepts. This process also encourages them to think about their activities as applying specific skills.

You can also give this reflection at the beginning of a week and use it as a weekly planning document. This method requires students to predict what will take up their time in the coming week.

Weekly Progress for: 5/8-5/12

Skills Used	Documentation	Explanation
Pick One to Two	Images/Video/Diagrams	Text/Audio/Video
Create a Plan		
Enact a Plan		
Learn a Skill		
Problem-Solving		
Prototyping/ Iteration		

WHAT IS MAKING?

The term *"making"* is so broad as to be undefinable. It is an "empty vessel," as I once heard Sylvia Libow Martinez describe it at a FabLearn conference. In marketing terms, an empty vessel is a company name with no defined meaning, allowing the brand to be defined in any way the company sees fit.

When I started my Master of Fine Arts program in the painting department, I saw second and third-year painters making 3D and 2D art in all artistic mediums. It became a joke to point at all sorts of art and ask, "Is that painting?" I loved paint and brushes and the process of making images on canvas. I was confident that I would be one of the few who spent my time refining and developing my painting process.

Three years later, my thesis exhibition included paintings, but also digital art, found-object sculpture, installation, and more. It had a gumball machine where you could buy birdseed to feed the pigeons of Philadelphia. It was built into a mobile storage container and included an audio tour. It's safe to say that it was not what I envisioned when I started, and I am profoundly grateful for the opportunity to expand my idea of art and making. Ultimately, I found that my medium is not paint but ideas.

I was in grad school a few years after the start of *Make:* magazine, and it was there I first heard the term *maker*. I thought at the time that all of us in the program should call ourselves makers instead of painters, ceramicists, and sculptors. It was more accurate to how the program worked and how we thought about art.

Traditionally, the term *painter* implies a discipline — the use of paint to create a thing called art. Enlarging my sense of what I did from painter to artist allowed me to envision more ways to turn my ideas into artistic products. Later, moving from artist to maker, I went one more level up the food chain. A visual artist is a maker who makes objects meant to be viewed from an aesthetic perspective, what we generally call art.

This is why maker education holds so much promise for schools. I can imagine all sorts of disciplines that fall under the maker umbrella. Writers, scientists, mathematicians, athletes, etc., can all be makers. I don't just mean that they are writers who also make *things*; I mean that the fundamental skills of the creative process apply to all these disciplines. When learning basic writing skills, we must imitate existing authors' methods and styles. Scientists designing experiments will need to understand and modify the natural world to test it. Mathematicians creating new models will need to innovate by combining existing knowledge in new ways.

The ideal maker education program not only teaches students these creative skills so they can go out and apply them to whatever problems they see in the world, but it also works directly with these disciplines within the school. The school makerspace should be a library, not a tech lab.

Making Versus STEM

STEM has been a popular buzzword for over a decade now. Before that, it was "21st-century skills." We can add art to the mix to create STEAM. Often the makerspace is placed firmly in the STEM/STEAM department of the school. Most schools consider it part of the technology and engineering curriculum. It may use physics and math concepts, but those are not taught directly in the makerspace.

I have stated several times in this book that engaging maker and design programs in elementary and middle school often turn into engineering and robotics classes in high school. I have nothing against engineering and robotics; all high schools should offer these valuable courses. But engineering and robotics are disciplines equivalent to painting and writing and should not be conflated with making.

Just as the skills of making can be applied to painting, cooking, and designing experiments, they can also be applied to engineering and robotics. Engineering involves imitation, modification, and innovation, and you could teach a rich robotics course using the three modes. Still, most students will not sign up for those courses if they are electives. Most high schools don't have required tech courses.

Makers are not, by default, engineers or techies. They are not required to be artists or crafters. They could be all these things or none. A maker uses their creative skills to turn their ideas into realities.

The perfect makerspace would include all tools that creative people may want. It would consist of an art studio, video editing equipment, digital fabrication, crafting, cooking, robotics, chemistry, and more. Such schools and spaces exist, but most children don't have access to them.

So in the interest of space and budget, we let the art studio keep the art supplies, and the chemistry stays in the science lab. The makerspace can hold the tools needed for making physical objects as efficiently as possible. The 3D printers and CNC machines can fabricate ideas almost like magic. Hand and power tools allow students to experience what it feels like to shape physical materials to their needs. Students can use electronics to perform various functions, allowing them to understand and control our digital world.

Many excellent books will tell you how all those tools work. There are makerspaces with classes you can take to gain experience. I hope this book gives you ideas and resources to teach the creative process in the makerspace. By teaching students in high school or college how to use the tools and how to use the creative process, we empower them to become creative problem solvers in all walks of life.

We Are All Makers

I have no sympathy with the belief that art is the restricted province of those who paint, sculpt, make music and verse. I hope we will come to an understanding that the material used is only incidental, that there is artist in every man; and that to him the possibility of development and of expression and the happiness of creation is as much a right and as much a duty to himself, as to any of those who work in the especially ticketed ways.
—Robert Henri, *The Art Spirit*, 1923

When I watch my young son imagining all sorts of weapons and tools out of sticks, or cutting cardboard into shields and ships, it's clear that the desire to make and create starts very early. When I look at the resources that some students have while others sit idle in large classes with few opportunities, it's just as clear there is a long way to go. I have worked at schools that value creativity and wanted students to have the opportunity to solve interesting problems. Vast resources or few, it is the will of the school community that is most critical to a program's success.

I believe that maker education has the power to change traditional education in this country for the better. Most reforms tinker around the edges of our school system, shifting the emphasis from one factor to another and ultimately coming back again. Schools that are truly different are few and far between. Maker education holds the potential to significantly enrich the experience of school for millions of students. It can bring meaningful and engaging learning to every content area at every age, because making is not a separate thing; it is a thing inherent to all areas of the school.

What's more, even though shops and studios have existed for centuries, makerspaces seem new. There are no expectations of what they are; there are no rigid standards and historical precedents. Parents don't complain when the makerspace doesn't teach it like they did "back in my day." Even the arts don't have that freedom in most schools.

But for maker education to fulfill this promise, it has to be meaningful, and as much as possible, it has to be intentional. There have to be methods to turn explicit knowledge into tacit knowledge, making it possible to discuss and reflect on what it means to make. Students need more than technical skills; they need to know how to pursue a project and what strategies will work for different types of problems.

If it is meaningful, our schools will support it. Vacant computer labs and wood shops will become makerspaces, and makerspace teachers will be looked at like librarians, helping everyone. The makerspace itself will be viewed as a community resource, not just a tech lab. In this way, the audience will be broad and diverse, and the invitation will be out there for every student to benefit from it as they see fit. They will learn how to imitate, modify, and innovate in the subjects that are most meaningful to them.

The tools exist, the ideas exist, we just need to turn those ideas into realities.

ACKNOWLEDGMENTS

I've written blogs, articles, and short pieces over the years, but always thought this topic could be a book. For helping to make that happen I want to thank Dale Dougherty, Kevin Toyama, Juliann Brown, Ann Martin Rolke, and all the people at Make: who contributed ideas and feedback. It's easier to do something big when you know you've got help.

I've been helped by a lot of great educators and administrators over the years. I owe a lot of my young creative self to Walt Bartman, who challenged me to make 100 paintings. I want to thank all the collaborative teachers at Noble Academy and Bullis for sharing their ideas and students with me. Linda Hale, Jamie Dickie, and Faith Darling all took chances on me to set up maker programs and I appreciate those opportunities. Thanks to Stacey Roshan for advice on writing about education while simultaneously being an educator.

Thanks to my family for cheering me on, especially my mom and dad for reading early drafts and giving me their advice and feedback. To my wife, Brienne, and my son, Emmett, thank you for giving me the love, encouragement, and time to write this book. This is for you.

ABOUT THE AUTHOR

Matt Zigler is a teacher, artist, and maker. He is the Bullis Innovation and Technology Lab (BITlab) Coordinator at Bullis School in Potomac, Maryland, where he works with students and teachers to bring design thinking and the maker process into traditional content area classes, teach maker-related classes, and oversees a state-of-the-art makerspace and Fab Lab.

Matt has presented on how to design and build a makerspace that meets the needs of specific schools, how to create a culture of making and innovation, and how to develop maker programming for all levels of ability. His art has been displayed at such places as the North Carolina Museum of Natural Sciences and the Rosenwald-Wolf Gallery at the University of the Arts. Matt lives in Rockville, Maryland.

Printed in the USA
CPSIA information can be obtained
at www.ICGtesting.com
JSHW051808040324
58548JS00012B/345

9 781680 457995